1 MONTH OF
FREE
READING

at

www.ForgottenBooks.com

By purchasing this book you are eligible for one month membership to ForgottenBooks.com, giving you unlimited access to our entire collection of over 1,000,000 titles via our web site and mobile apps.

To claim your free month visit:

www.forgottenbooks.com/free897737

ISBN 978-0-266-84257-6
PIBN 10897737

This book is a reproduction of an important historical work. Forgotten Books uses
state-of-the-art technology to digitally reconstruct the work, preserving the original format
whilst repairing imperfections present in the aged copy. In rare cases, an imperfection in
the original, such as a blemish or missing page, may be replicated in our edition. We do,
however, repair the vast majority of imperfections successfully; any imperfections that
remain are intentionally left to preserve the state of such historical works.

MARYLAND GEOLOGICAL SURVEY

TALBOT COUNTY

MARYLAND

GEOLOGICAL SURVEY

TALBOT COUNTY

BALTIMORE
JOHNS HOPKINS PRESS
1 9 2 6

PRESS OF MEYER & THALHEIMER
BALTIMORE, MD.

ADVISORY COUNCIL

RAYMOND A. PEARSON - - - - - EXECUTIVE OFFICER
PRESIDENT UNIVERSITY OF MARYLAND

FRANK J. GOODNOW - - - - - - Ex-OFFICIO MEMBER
PRESIDENT JOHNS HOPKINS UNIVERSITY

ROBERT W. WILLIAMS - - - - - - - BALTIMORE

JOHN B. FERGUSON - - - . - - - - HAGERSTOWN

SCIENTIFIC STAFF

LETTER OF TRANSMITTAL

To His Excellency ALBERT C. RITCHIE, *Governor of Maryland,*

Sir:—I have the honor to present herewith a report on The Physical Features of Talbot County. This volume is the tenth of a series of reports on the county resources, and is accompanied by large scale topographical, geological, and agricultural soil maps. The information contained in this volume will prove of both economic and educational value to the residents of Talbot County as well as to those who may desire information regarding this section of the State. I am,

Very respectfully,

EDWARD BENNETT MATHEWS,
State Geologist.

JOHNS HOPKINS UNIVERSITY,

BALTIMORE, *September, 1926.*

CONTENTS

CONTENTS

PAGE

CONTENTS

ILLUSTRATIONS

PREFACE

This volume is the tenth of a series of reports dealing with the physical features of the several counties of Maryland.

The *Introduction* contains a brief statement regarding the location and boundaries of Talbot County together with its chief physical characteristics.

The Physiography of Talbot County, by Benjamin L. Miller, comprises a discussion of the surface characteristics of the county, together with a description both of the topographic forms and of the agencies which have produced them.

The Geology of Talbot County, by Benjamin L. Miller, deals with the stratigraphy and structure of the county. Many stratigraphical details are presented, accompanied by local sections.

The Mineral Resources of Talbot County, by Benjamin L. Miller, deals with the economic possibilities of the various geological deposits of the county. Those which have been hitherto employed are fully discussed, and suggestions are made regarding the employment of others not yet utilized.

The Soils of Talbot County, by Hugh W. Bennett, W. E. Tharp, W. S. Lyman, and H. L. Westover, contains a discussion of the leading soil types of the county and their relation to the several geological formations. This investigation was conducted under the direct supervision of Professor Milton Whitney, Director of the Bureau of Soils of the U. S. Department of Agriculture.

The Climate of Talbot County, by Roscoe Nunn, is an important contribution to the study of the climatic features of the county. Mr. Nunn is Section Director in Baltimore of the U. S. Weather Bureau, and also Meteorologist of the Maryland State Weather Service.

The Hydrography of Talbot County, by B. D. Wood, gives a brief account of the water supply of the county, which, as in the case of the other Coastal Plain counties, affords but little power for commercial purposes.

The Magnetic Declination in Talbot County, by L. A. Bauer, contains much important information for the local surveyors of the county. Dr. Bauer has been in charge of the magnetic investigations since the organization of the Survey and has already published two important general reports upon this subject. He is the Director of the Department of International Research in Terrestrial Magnetism of the Carnegie Institution.

The Forests of Talbot County, by F. W. Besley, is an important contribution and should prove of value in the further development of the forestry interests of the county. Mr. Besley is State Forester of Maryland.

The State Geological Survey desires to extend its thanks to the several national organizations which have liberally aided it in the preparation of several of the papers contained in this volume. The Director of the U. S. Geological Survey, the Chief of the U. S. Weather Bureau, the Chief of the U. S. Bureau of Soils of the U. S. Department of Agriculture have granted many facilities for the conduct of the several investigations and the value of the report has been much enhanced thereby.

THE

PHYSICAL FEATURES

OF

TALBOT COUNTY

THE PHYSICAL FEATURES OF TALBOT COUNTY

INTRODUCTION

Talbot County lies to the east of Chesapeake Bay and forms a part of what is known as the "Eastern Shore" of Maryland. It is included between the parallels of 38° 34' and 38° 57' north latitude and between the meridians of 75° 54' and 76° 25' west longitude. It has an area of 290.83 square miles, making it one of the smaller counties of the state in geographic extent. In population and wealth, however, it far exceeds several of the larger counties.

Next to Kent, Talbot County is the oldest of the Eastern Shore counties of the State having been founded about 1661. It was named for Grace Talbot, daughter of George, first Lord Baltimore. Originally the county embraced much more territory but it was gradually restricted by the erection of the adjoining counties of Queen Anne's, Dorchester, and Caroline.

Talbot County lies entirely within the Atlantic Coastal Plain and its physical features are characteristic of that physiographic province. It is almost entirely surrounded by tide-water estuaries and bays, many of which extend into the county for many miles. The greatest elevations in the county are little more than 70 feet above sea level while almost one-half the county is less than 25 feet above sea level.

The county is well provided with transportation facilities as two lines of railroad, the Pennsylvania and the Baltimore, Chesapeake, and Atlantic, extend entirely across the county while two steamboat lines from Baltimore make regular trips to various wharves on the Choptank and Tred Avon rivers and to Claiborne on the Bay shore. Besides, sailing vessels pass up many of the other tidewater streams during the seasons when fruit, vegetables, and grain are being shipped to market and a portion of the produce of the country is transported in such vessels.

Farming is the chief occupation of the county and almost all of the land is under cultivation. The many well-kept farms and attractive farm buildings that are to be seen in all parts of the county indicate the prosperity of the inhabitants. The waters which wash the shores of the county abound in oysters, fish, and crabs; and the capture and care of these furnish occupation to many of the people. Within recent years the attractions of Talbot County have drawn many people to its borders during the summer season and many summer homes and hotels have been erected along the shores to accommodate those who delight in boating, bathing, and fishing.

Easton, the county seat and the largest town in the county, is a prosperous enterprising town with a population of about 4,000. It is centrally located and easily accessible to all parts of the county as two railroads pass through it while a steamboat line running between Baltimore and points on the Choptank River comes within one mile of the town. Oxford, St. Michaels, Trappe, and Tilghman are the other important towns, and small villages are plentifully distributed throughout the county.

DEVELOPMENT OF KNOWLEDGE CONCERNING THE PHYSICAL FEATURES OF TALBOT COUNTY

BY

BENJAMIN L. MILLER

The location of Talbot County on navigable waters and its early settlement are together responsible for it being mentioned in many publications extending over a period of about 300 years. Naturally the first references are somewhat vague and consist merely of geographic notes or of maps on which the geographic features of the region are portrayed. For this reason it is considered advisable to divide the discussion of the literature references into two classes: those which are primarily of geographic interest and those which pertain to the description of the geologic features of the region.

THE HISTORY OF GEOGRAPHIC RESEARCH.

Chesapeake Bay was probably entered and explored by Ayllon, a Spaniard, in 1526 and perhaps by Gomez, another Spaniard, a few years later. The maps prepared as the result of these voyages are vague and general but there are some who believe that an indentation in the ocean shore is supposed to represent Chesapeake Bay. Until the English settlement at Jamestown these maps were accepted as authoritative and were reproduced in the works of several different writers.

In the summer of 1608, one year after the establishment of the English colony at Jamestown, Captain John Smith with fourteen companions started out to explore Chesapeake Bay. They went up the Bay as far as the Patapsco River and then returned to Jamestown, having been away less than three weeks. Later in the summer Smith again headed another party for the same purpose and that time went to the head of the Bay. His map and also his notes indicate that he made no close-range observations along the Eastern Shore between the Nanticoke and Sassafras rivers as he kept close to the western shore in that portion of the Bay. However, he attempted to delineate the characteristics of the entire Bay on his map. He made the mistake, that one might naturally make in taking observations at that great distance, of supposing that three large islands occurred in that portion of the Bay. These he named the "Winstone Islands." They evidently represent Kent Island, the peninsula of Talbot County lying between the Miles and Choptank rivers, and the western portion of Dorchester County.

Smith's map was published in England in 1612 and served as the source of information for almost all the maps of this region which were published during the next fifty years. The Lord Baltimore map, published in 1635, reproduces the same errors made by Smith in delineating the outlines of the shores. The map is interesting in that conical-shaped hills, or mountains, are represented in the area now included in Queen Anne's, Talbot, and Caroline counties.

A map drawn by Virginia Farrer and published in 1651 is greatly distorted and is altogether less reliable than some of the earlier maps, yet it crudely represents the Choptank and Chester rivers.

In 1666 a fifth map of Maryland, by George Alsop, was published. This is interesting in that it represents the Eastern Shore to be cut into a number of peninsulas of approximately the same size and shape by the Choptank, Wye, Chester, "Sassafrix," "Elke," and two other unnamed rivers that all flow almost due west, are of equal size and empty in "——piacke Bay" along a straight north and south line.

In 1660 Augustine Herman offered to make a map of Maryland in return for a manor along Bohemia River and on the acceptance of the offer by Lord Baltimore moved his family to Maryland and began the work. He was engaged in the work for ten years and it was not until 1673 that the map was finally published. The map is

FIG. 1.—VIEW OF BLUFFS NEAR DOVER BRIDGE ON CHOPTANK RIVER.

FIG. 2.—VIEW SHOWING FARM LANDS ON THE TALBOT PLAIN IN TALBOT COUNTY.

remarkably good and for the first time the configuration of the Eastern Shore counties was correctly shown. Most of the names which appear on the map are those still in use. "Talbot" is written on that portion of territory lying between the Chester and "Choptanck" rivers and extending from "The Great Bay of Cheseapeake" to "Delawar Bay." Just as Smith's map served as the basis for later maps for many years so did Herman's map appear in many volumes and in many forms during the next century.

In 1735 Walter Haxton in a "Mapp of the Bay of Chesepeacke" added many details that were lacking from Herman's map. On this map nearly all the points and inlets along the Bay shore were correctly shown and many observations on the depths of the water are given as the map was intended for the use of mariners.

After Haxton's map appeared there was little to be added so far as the Eastern Shore was concerned and the only other maps worthy of mention here are the maps by Dennis Griffith in 1794, by J. H. Alexander 1834-1840, and by S. J. Martinet in 1865. These contained much new information on the political boundaries, towns, and roads but in the representation of natural features were practically copies of the earlier maps.

The last map to be mentioned is the one accompanying this report and which has been prepared by the United States Geological Survey in coöperation with the Maryland Geological Survey. The details shown on this map, when compared with Smith's map, furnish decisive evidence of the continual progressive development that the Eastern Shore has undergone during the past 300 years, as well as showing the great improvements which have been made in the art of map making and map publishing.

THE HISTORY OF GEOLOGIC RESEARCH

GENERAL.

The first paper on the Geology of North America worthy of especial consideration was published by William Maclure in 1809 and republished in more complete form in 1817, 1818, and 1826. In this report and on the accompanying map all the Atlantic Coastal Plain was supposed to be composed of a single geologic unit which was called the "Alluvial Formation." It received the name from the un-consolidated condition of the materials which suggested to him an alluvial origin. The boundary between the Coastal Plain and the Piedmont Plateau or between what he called the "Alluvial" and "Primitive" formations, was drawn with a high degree of accuracy. This first contribution by Maclure, although extremely general, nevertheless possesses great value as it was the impetus needed to stimulate systematic geologic investigation.

In 1820 Hayden objected to Maclure's explanation of the origin of the Coastal Plain materials and stated that they were not alluvial in character but marine and were brought to their present position by an ocean current that swept over the eastern part of the country.

THE TERTIARY FORMATIONS.

The first attempt to differentiate the various formations of which we now know the Coastal Plain to be composed were made in 1824 by John Finch, an Englishman, who traveled in this country during the preceding year. He recognized the presence of Secondary (Cretaceous) and Tertiary formations which he provisionally cor-related with similar formations in Europe. Some of the correla-tions which he made were incorrect yet his contribution was of great value in that he showed that it would be possible to divide the beds of unconsolidated materials into several formations on the basis of the fossils which they contained.

T. A. Conrad was the first paleontologist of importance to en-gage in the investigation of the Tertiary fossils of Maryland. He published his first article on this subject in 1830 and two years later, as a result of his further paleontologic data, he divided the Coastal Plain deposits into six formations. Conrad continued to study Coastal Plain fossils and numerous articles from his pen were published from 1830 to 1867. To him much credit is due for the accurate determination of the major groups of strata composing the Coastal Plain and without Conrad's careful determinations and descriptions of the contained fossils, it would have been impossible to make much progress. He divided the Tertiary into the Eocene and Miocene although Isaac Lea was the first person to apply the term "Eocene" to any deposits in this country.

The first Geological Survey of Maryland, under the direction of Ducatel, the State Geologist, and Alexander, the State Engineer, contributed much to our understanding of the geology of the State and their reports issued from 1834 to 1842 contain the results of careful, systematic investigations prosecuted in all parts of the State.

Several writers between 1830 and 1860 emphasized the value of the shell and greensand marls of the Coastal Plain for fertilizing purposes and their contributions contain much information in regard to the Tertiary deposits of this region. They seem to have exaggerated the value of the marls yet there can be no question but that the marls do possess the requisite properties for improving many soils. Pierce, Purvis, Ruffin, Higgins, and Tyson were the men who were most active in recommending to farmers that they make use of the easily accessible marl deposits, while in the adjoining states of Delaware and Virginia, Booth and W. B. Rogers were equally active in urging similar action on the residents of their states.

In the further study of the Tertiary deposits the work has been done almost entirely by Heilprin and members of the Maryland Geological Survey and United States Geological Survey. The contributions of Clark, Dall, Shattuck, and Martin are of the greatest importance and due to these men we have the present divisions and classification used in this report.

THE QUATERNARY FORMATIONS.

In general, the early geologic investigation either entirely ig-. nored the surficial deposits or else merely referred to them incidentally. Chester in 1884 and 1885 described the surface gravels of the Eastern Shore and gave an explanation of their origin. McGee, however, was the first one to make systematic observations of these materials over extended areas of the Coastal Plain and in several papers published during the years 1886 to 1891 we have much valuable information. He proposed the name "Columbia" for the Quaternary deposits and divided them into fluviatile and interfluviatile phases which he considered contemporaneous. Although McGee's deductions are not all accepted now yet his observations were carefully made and served as the basis for later work.

Darton continued McGee's work and early divided the surficial Quaternary deposits into two formations which he designated "Earlier Columbia" and "Later Columbia." The latest important contributions have been made by Shattuck, who in 1901 divided the Columbia deposits into three formations which he named the "Sunderland," "Wicomico," and "Talbot." This is the classification that is used in this report.

BIBLIOGRAPHY

1612.

SMITH, JOHN. A Map of Virginia with a Description of the Country, the Commodities, People, Government, and Relegeon. Written by Captaine Smith, sometime Governour of the Countrey. Oxford, printed by Joseph Barnes, 1612. 4 to, 174 pp.

The map imperfectly represents the Bay shore features of Talbot County.

1624.

SMITH, JOHN. A Generall Historie of Virginia, New England, and the Summer Isles, etc. London, 1624. (Several editions.)

This work contains many interesting notes on the physiography of Chesapeake Bay and its tributaries, and briefly described the clays and gravels along their shores. For a reproduction and discussion of Smith's map see Md. Geol. Surv., Vol. II, pp. 347-360.

1635.

ANON. A Relation of Maryland; Together With a Map of the Countrey, The Conditions of Plantation, His Majestie's Charter to the Lord Baltimore, translated into English. London, 1635.

(Repub.) Sabine's Reprints, 4 to., ser., No. 2, New York, 1865, pp. 1-65, with appendix pp. 67-73.

The map accompanying this report is similar to that made by Smith but, in addition, conical hills or mountains are represented in the area now included in the Eastern Shore counties of Queen Anne's, Talbot, and Caroline.

1651.

FARRER, VIRGINIA. A mapp of Virginia discovered to ye Hills, and in it's Latt: From 35 deg: & ½ neer Florida, to 41 deg: bounds of New England. John Goddard sculp. Domina Farrer Collegit. Are sold by I. Stephenson at ye Sunn below Ludgate: 1651.

Includes a greatly distorted representation of the Chesapeake Bay region.

1666.

ALSOP, GEORGE. A Character of the Province of Maryland.

(Repub.) Gowan's Bibliotheca Americana, New York, 1869, No. 5.

Contains a generalized map in which the Eastern Shore rivers and peninsulas are represented as being very regular.

1673.

HERMAN, AUGUSTINE. Virginia and Maryland As it is Planted and Inhabited this present year 1670.

This is a very good map of the two colonies and the first accurate one of the Chesapeake Bay region ever prepared.

1735.

HAXTON, WALTER. To the Merchants of London Trading to Virginia and Maryland. This mapp of the Bay of Chesapeack with the Rivers Potomack, Patapsco, North East and part of Chester, Is humbly dedicated & presented by Walter Haxton 1735.

The map correctly represents most of the topographic features of the Bay shores.

1817.

MACLURE, WM. Observations on the Geology of the United States of America, with some remarks on the effect produced on the nature and fertility of soils by the decomposition of the different classes of rocks. 12 mo., 2 pls., Phila., 1817.

Is an elaboration of an article published in 1809 in Trans. Amer. Phil. Soc., O. S., Vol. VI, pp. 411-428. Repub. in Trans. Amer. Phil. Soc., N. S., Vol. I, 1818, 191 pp.

This work is classic as it was the first attempt to treat the geology of the entire country and it contains the first published geological map of the United States. The whole Coastal Plain constitutes the "Alluvial" formation and the Piedmont Plateau, the "Primitive."

1818.

MITCHELL, SAMUEL L. Essay on the Theory of the Earth by M. Cuvier to which are now added Observations on the Geology of North America by Samuel L. Mitchell. 8 vo., 431 pp., 8 pls. New York, 1818.

He believes that the fossil remains in this region "afford proofs.......of a deposite from inland floods since the oceanic strata were formed." p. 395.

1820.

HAYDEN, HORACE H. Geological Essay; or An Inquiry into some of the Geological Phenomena to be found in various parts of America, and elsewhere. 8 vo., 412 pp. Baltimore, 1820.

The writer contends that the unconsolidated deposits bordering the Atlantic Ocean are not alluvial materials but have been brought to their present position by an ocean current that swept over the eastern part of the country in a southwesterly direction. The rise of the ocean is believed to have been caused by an increase of water due to the melting of the polar ice produced by a shifting of the earth's axis.

1824.

FINCH, JOHN. Geological Essay on the Tertiary Formations in America.

(Read before Acad. Nat. Sci., Phila., July 15, 1823.) Amer. Jour. Sci., Vol. VII, pp. 31-43, 1824.

Objection is made to the term "Alluvial formation" of Maclure and others on the ground that the deposits are for the most part not of alluvial origin and also as used, the term includes a number of distinct formations that can be correlated with the "newer secondary and tertiary formations of France, England, Spain, Germany, Italy, Hungary, Poland, Iceland, Egypt. and Hindoostan." The writer makes some provisional correlations with European formations which are now known to be incorrect. He admits, however, that the data at his disposal are insufficient for accurate correlation.

1826.

PIERCE, JAMES. Practical remarks on the shell marl region of the eastern parts of Virginia and Maryland, etc; extracted from a letter to the Editor.

Amer. Jour. Sci., Vol. XI, pp. 54-59, 1826.

Mentions the occurrence of shell marl of marine origin in the "alluvial" district of Maryland on both sides of Chesapeake Bay and discusses its value as a fertilizer in the renovation of exhausted soils.

1834.

DUCATEL, J. T. and ALEXANDER, J. H. Report on the Projected Survey of the State of Maryland, pursuant to a resolution of the General Assembly. 8 vo. 39 pp. Annapolis, 1834. Map. Several editions.

Amer. Jour. Sci., Vol. XXVII, 1835, pp. 1-39.

"Two varieties of shell marl, one composed principally of clam shells imbedded in clay; the other consisting of pectens (scallop shells) enveloped by an indurated ferruginous clay" are reported to occur on Corsica Chreek.

1835.

CONRAD, T. A. Observations on the Tertiary Strata of the United States.

Amer. Jour. Sci., Vol. XXVIII, 1835, pp. 104-111, 280-282.

He considers the Miocene absent in this region, the Older Pliocene resting directly upon the Eocene. The beds containing *Perna maxillata* are referred to the Older Pliocene and the St. Mary's river beds to the Medial Pliocene.

DUGATEL, J. T. and ALEXANDER, J. H. Report on the New Map of Maryland, 1834. Annapolis, 1835(?). 8 vo. 591 pp. Two maps and one folded table. Contains Engineer's and Geologist's Reports which were also issued separately. Md. House of Delegates, Dec. Sess. 1834.

Ducatel says that he believes the shell marl deposit underlies the Eastern Shore but is not exposed south of the Choptank River. He gives the dip as 5° to the southwest. He also says that the surface of the marl undulates. He describes deposits of shell marl at the head of Southeast Creek in the vicinity of Church Hill, on the northeast side of Corsica Creek, at many places on the southwest side of Corsica Creek extending to the head of the branch south of Centreville, at the head of Reed's Creek, on Back Wye River, and on Chew's Island. Sixteen analyses of these marls are given. The soils of the county are described and the value of the shell marls as fertilizers is discussed as well as methods of working and applying the material. Bog iron ore of good quality is reported at the head of Hamilton and South East creeks. A chalybeate spring is said to be located on the farm of Mr. Levi Paccault near Wye Mills.

1836.

DUCATEL, J. T. and ALEXANDER, J. H. Report on the new Map of Maryland, 1835. 8 vo. 84 pp. Maps. Annapolis, 1835.

Md. Pub. Doc., Dec., Sess., 1835. Engineer's Report pp. 1-34, Geologist's Report, pp. 35-85.
Both reports also published separately.

PURVIS, M. On the Use of Lime as a Manure.

Translated for the Farmer's Register, Shellbanks, Va., 1835. Reviewed in Amer. Jour. Sci., Vol. XXX, 1836, pp. 138-163.
Mention is made of the use of greensand as a fertilizer.

1837.

DUCATEL, J. T. Outline of the Physical Geography of Maryland, embracing its prominent Geological features.

Trans. Md. Acad. Sci. and Lit., Vol. I, Pt. I, 1837, pp. 24-55 with map.

A general description of the distribution and characteristics of the Miocene of the with many details of local features. It is a general summary of information pre-viously published in various places. Mention is made of the covering of boulders and coarse gravel near the inner edge of the Secondary (Cretaceous) rocks while farther out the sands and clays of the Secondary and Tertiary formations are uncovered. The whole county is said to be underlain by Tertiary deposits though no reference is made to any particular locality in the county.

1838.

CONRAD, T. A. Fossils of the Medial Tertiary of the United States. No. 1, 1838.

[Description on cover: 1839 & '40], 32 pp., pls. I-XVII. Repub. by Wm. H. Dall, Washington, 1893.
A general description of the physiography and geology of the entire state is given Atlantic Coastal Plain is given. The Miocene is called the Medial Tertiary or Older Pliocene and the Eocene is called Lower Tertiary.

1840.

CONRAD, T. A. Fossils of the Medial Tertiary of the United States. No. 2, 1840.

[Description on cover: 1840-1842], pp. 33-56, pls. XVIII-XXIX. Repub. by Wm. H. Dall, Washington, 1893.
Astarte cuneiformis from Wye Mills is described and figured.

1841.

VANUXEM, LARDNER. On the Ancient Oyster Shell Deposits observed near the Atlantic Coast of the United States. (Read April 7, 1841.)

Proc. Assoc. Amer. Geol. and Nat., pp. 21-23.

The writer agrees with Ducatel in the view that the deposits were made by the Indians, though he admits that Conrad has some evidence to prove that they were formed by natural agencies. He believes that possibly the deposit at Easton is due to natural causes as the valves seem to be together there, while elsewhere they are separated.

1842.

CONRAD, T. A. Observations on a portion of the Atlantic Tertiary Region, with a description of new species of organic remains. 2d Bull. Proc. Nat. Inst. Prom. Sci., 1842. pp. 171-192.

The Miocene and Eocene are said to not be connected by a single fossil common to both periods while three forms found in the Upper Secondary are found in the Eocene. The Medial Tertiary (Miocene) is said to appear near Chestertown and Wye Mills.

1843.

DUCATEL, J. T. Physical History of Maryland. Abstract, Proc. Amer. Phil. Soc., Vol. III, 1843, pp. 157-158.

"The Eastern Shore is shown to consist of something more than arid sand-hills and pestilential marshes; and the Western Shore not to depend exclusively upon the rich valleys of Frederick and Hagerstown for its supplies."

1850.

HIGGINS, JAMES. Report of James Higgins, M. D., State Agricultural Chemist, to the House of Delegates. 8 vo. 92 pp. Annapolis, 1850.

Contains detailed descriptions and many analyses of the various kinds of soils found on the Eastern Shore of Maryland. The greensand and shell marl deposits of the counties lying north of the Choptank River are discussed at length. Reference is made to several localities in Talbot County where they occur.

1852.

FISHER, R. S. Gazateer of the State of Maryland compiled from the returns of the Seventh Census of the United States. New York and Baltimore, 8 vo. 1852, 122 pp.

Contains numerous descriptions of the geography and geology of the various portions of the State.

1860.

TYSON, PHILIP T. First Report of Philip T. Tyson, State Agricultural Chemist, to the House of Delegates of Maryland. January,

1860. 8 vo. 145 pp. Maps. Appendix. Mineral Resources of Md. 20 pp. Annapolis, 1860.

1862.

TYSON, PHILIP T. Second Report of Philip T. Tyson, State Agricultural Chemist, to the House of Delegates of Maryland. January, 1862. 8 vo. 92 pp., Annapolis, 1862.

1867.

HIGGINS, JAMES. A Succinct Exposition of the Industrial Resources and Agricultural Advantages of the State of Maryland. 8 vo., 109+III pp.

Md. House of Delegates, Jan. Sess., 1867, (DD).

Md. Sen. Doc., Jan. Sess., 1867, (U).

Contains descriptions of the soils and physiographic features of each of the counties of the State.

1880.

HEILPRIN, ANGELO. On the Stratigraphical Evidence Afforded by the Tertiary Fossils of the Peninsula of Maryland.

Proc. Acad. Nat. Sci., Phila., Vol. XXXII, 1880, pp. 20-33.

After a careful examination of the fossils found along the Patuxent, Choptank, and St. Mary's rivers and the Calvert Cliffs, the author proposes the separation of the Miocene into the Older and Newer periods. The beds at Fair Haven are typical Older Miocene and the St. Mary's lower Patuxent and Choptank river beds belong to the Newer Miocene.

1882.

——————————On the relative ages and classification of the Post Eocene Tertiary Deposits of the Atlantic Slope.

Proc. Acad. Nat. Sci., Phila., Vol. XXXIV, 1882, pp. 150-186.

Asbtract: Amer. Jour. Sci., 3d ser., Vol. XXIV, 1882, pp. 228-229.

Amer. Nat., Vol. XVII, 1883, p. 308.

From a comparison of faunas the Eocene deposits of Maryland are correlated with the Eo-Lignitic of Alabama, and the Miocene beds of the State are grouped in a division called the Marylandian which is supposed to be older than any other Miocene beds of this country, with the possible exception of the basal Miocene beds of Virginia which may be contemporaneous.

1884.

CHESTER, FREDERICK D. The quaternary Gravels of Northern Delaware and Eastern Maryland, with map.

Amer. Jour. Sci., 3d ser., Vol. XXVII, 1884, pp. 189-199.

The author believes that the peninsula of Eastern Maryland and Delaware was covered with gravels, clay and sand brought down by the Delaware River during the Ice Age and deposited in an estuary.

HEILPRIN, ANGELO. The Tertiary Geology of the Eastern and Southern United States.

Jour. Acad. Nat. Sci., Phila., Vol. IX, pt. 1, pp. 115-154, map. 1884.

The distribution of the Tertiary strata of the State is given approximately. The Eocene is correlated with the base of the Buhrstone or the Eo-lignitic of Alabama and with the London Clay. The Miocene of the State is divided into two formations, the older or Marylandian which is regarded as possibly Oligocene is age, and the newer or Virginian. The former is exposed in Anne Arundel, Calvert, and Charles counties and the latter at Easton, on the Choptank River and in St. Mary's County.

————————————————Contributions to the Tertiary Geology and Paleontology of the United States. 4 to, 117 pp. 1 map. Phila., 1884.

Contains a number of articles all but one of which was previously published in the Proceedings or Jour. of the Philadelphia Academy of Sciences. Some of these articles are listed on preceding pages.

1885.

CHESTER, FREDERICK D. The gravels of the Southern Delaware Peninsula.

Amer. Jour. Sci., 3d ser., Vol. XXIX, 1885, pp. 36-44.

The gravels, sands, and clays of the entire peninsula of Eastern Maryland and Delaware are said to have been brought down by the Delaware River and spread out by estuaries and marine currents. In the northern part the materials were deposited in an estuary but in the southern part in the open ocean. Boulders carried by icebergs are found throughout the entire area, some of which are of large size.

1888.

McGEE, W. J. Three Formations of the Middle Atlantic Slope.

Amer. Jour. Sci., 3d ser., Vol. XXXV, 1888, pp. 120-143, 328-331, 367-388, 448-466, plate II.

The three formations discussed are the Potomac, (now divided into four formations), the Appomattox (Lafayette, later Brandywine), and the Columbia, (now divided into three formations). These are described in far greater detail than had ever been done before and the conclusions reached vary little from the views held at the present time.

——————————————————The Geology of the Head of Chesapeake Bay.

7th An. Report U. S. Geol. Surv., Washington, 1888, pp. 537-646.

(Abst.) Amer. Geol., Vol. I, 1887, pp. 113-115.

Contains a general discussion of the Pleistocene deposits of the State.

——————————————————The Columbia Formation.

Proc. Amer. Assoc. Adv. Sci., Vol. XXXVI, 1888, pp. 221-222.

The Columbia formation overlying unconformably the Cretaceous and Tertiary deposits of the Atlantic Coastal Plain is said to consist of series of deltas and terraced littoral deposits. It is said to pass under the terminal moraine to the northward. The Columbia materials are supposed to have been laid down during a period of glaciation long preceding the glacial epoch during which time the terminal moraine was formed.

DARTON, N. H.   Mesozoic and Cenozoic Formations of Eastern Virginia and Maryland.

Bull. Geol. Soc. Amer., Vol. II, 1891, pp. 431-450, map, sections.

(Abst.) Amer. Geol., Vol. VII, 1891, p. 185; Amer. Nat., Vol. XXV, 1891, p. 658.

Contains a description of the Potomac, Severn (marine Cretaceous), Pamunkey (Eocene), Chesapeake (Miocene), and Appomattox (Brandywine) formations as known at that time.

McGEE, W. J.   The Lafayette Formation.

12th Ann. Rept. U. S. Geol. Survey, pt. I, 1890-91, Washington, 1891, pp. 347-521.

The general characteristics of the entire Coastal Plain and of each of the formations composing it are discussed at length.

1892.

CLARK, WM. BULLOCK.   The Surface Configuration of Maryland.

Monthly Rept. Md. State Weather Service, Vol. II, 1892, pp. 85-89.

Contains a general summary of the physical features of the State.

DALL, WM. H. and HARRIS, G. D.   Correlation Papers: Neocene.

Bull. U. S. Geol. Survey No. 84, 1892, 349 pp., 3 maps, 43 figs.

House Misc. Doc., 52d Congress, 1st sess., Vol. XLIII, No. 337.

Contains a full discussion of all the literature of the Miocene and Pliocene of the United States published up to that time. Tentative correlations are made.

SCHARF, J. THOMAS. The Natural Resources and Advantages of Maryland, being a complete description of all of the counties of the State and the City of Baltimore. Annapolis, 1892.

Contains much general information concerning each county of the State.

1893.

CLARK, WM. BULLOCK. Physical Features (of Maryland).

Maryland, its Resources, Industries, and Institutions. Baltimore, 1893, pp. 11-54.

Contains short descriptions of the topography, climate, water supply, and water resources of the State.

WHITNEY, MILTON. The Soils of Maryland.

Md. Agri. Exper. Station, Bull. No. 21, 58 pp., map. College Park, 1893.

The principal soils of the State are described and their adaptability to different kinds of crops are discussed. A map is given showing their general distribution.

————————————————Description of the Principal Soil Formations of the State (Maryland).

Maryland, its Resources, Industries, and Institutions. Baltimore, 1893, pp. 181-211.

The writer describes the soils of the State, gives their distribution, and discusses their origin and adaptability.

WILLIAMS, G. H. and CLARK, WM. BULLOCK. Geology of Maryland.

Marlyand, its Resources, Industries, and Institutions. Baltimore, 1893, pp. 55-89.

The different geological formations of the State are briefly discussed.

1894.

CLARK, WM. BULLOCK. The Climatology and Physical Features of Maryland.

First Bien. Rept. Md. State Weather Service for years 1892 and 1893. Baltimore, 1894.

Contains a general discussion of the topography, geology, soils, and climate of the State.

DARTON, N. H. Artesian Well Prospects in Eastern Virginia, Maryland, and Delaware.

Trans. Amer. Inst. Min. Eng., Vol. XXIV, 1894, pp. 372-396, pls. 1 and 2.

Contains a general description of the Atlantic Coastal Plain formations with records of some of the important artesian wells of the region and a discussion of artesian water conditions throughout the area.

————————————————Outline of Cenozoic History of a Portion of the Middle Atlantic Slope.

Jour Geol., Vol. II, 1894, pp. 568-587.

Contains a description of the formations of the Atlantic Coastal Plain and a resumé of the geologic history of the region.

DARTON, N. H. Artesian Well Prospects in the Atlantic Coastal Plain Region.

Bull. U. S. Geol. Survey No. 138, 1896, 232 pp., 19 pls.

Contains a brief description of the Coastal Plain formations of the State with a discussion of their water-bearing properties. Records are given of many deep wells.

1897.

CLARK, WM. BULLOCK. Outline of the Recent Knowledge of the Physical Features of Maryland, embracing an Account of the Physiography, Geology, and Mineral Resources.

Md. Geol. Survey, Vol. I, 1897, pp. 141-228, pls. 6-13.

Contains a description of each of the geologic formations of the State recognized at that time.

ABBE, CLEVELAND, JR. General Report of the Physiography of Maryland.

Md. Weather Service, Vol. I, pp. 41-216, pls. 3-19, figs. 1-20. Baltimore, 1899.

Contains a full description of the physiographic features of the State.

1900.

————————————The Physiographic Features of Maryland.

Bull. Amer. Bur. Geog., Vol. I, pp. 151-157, 242-248, 342-355, 2 figs. 1900.

A concise description of the important physical features of each of the three physiographic provinces of the State.

1901.

CLARK, WM. BULLOCK. Maryland and its Natural Resources.

Official publication of the Maryland Commissioners to the Pan-American Exposition. 38 pp., map. Figs. Baltimore, 1901.

A brief account of the physical features and economic resources of the State.

SHATTUCK, GEORGE BURBANK. The Pleistocene Problem of the North Atlantic Coastal Plain.

Johns Hopkins Univ. Circ., Vol. XX, 1901, pp. 69-75; Amer. Geol., Vol. 28, 1901, pp. 87-107.

The views of McGee, Darton, and Salisbury concerning the Pleistocene deposits are summarized and compared with the writer's views. The wave-built terrace deposits are referred to four different formations, the Talbot, Wicomico, Sunderland, and Lafayette, the first three of which constitute the Columbia group. These formations are said to be separated by erosional unconformities.

1902.

DARTON, N. H. Preliminary List of Deep Borings in the United States. Part I, Alabama-Montana.

U. S. Geol. Survey, Water-Supply and Irrigation Paper No. 57. 60 pp. 1902.

RIES, HEINRICH. Report on the Clays of Maryland.

Md. Geol. Survey, Vol. IV, 1902, pp. 203-505, pls. 19-69.

Contains full description of the clay deposits and clay industries of the State.

1903.

————————————The Clays of the United States East of Mississippi River.

U. S. Geol. Survey, Prof. Paper No. 11, 1903, pp. 134-149.

1904.

MARTIN, G. C. et al Systematic Paleontology of the Miocene Deposits of Maryland.

Md. Geol. Survey, Miocene, 1904, pp. 1-508 pls. 10-135.

Contains descriptions and illustrations of all Miocene fossils recognized in Maryland up to that time. Many forms from Talbot County are included.

1906.

CLARK, WM. BULLOCK and MATHEWS, E. B. with the collaboration of others.

Md. Geol. Survey, Vol. VI, pp. 27-259, pls. 1-23, figs. 1-18, 1906.

Contains a full account of the physical features, geologic formations, and mineral products of the entire State.

MATHEWS, EDWARD B. The Counties of Maryland; their Origin, Boundaries, and Election Districts.

Md. Geol. Survey, Vol. VI, pp. 417-572, pls. 36-57, 1906.

Gives much valuable information concerning the boundaries of each of the counties of the State.

SHATTUCK, GEORGE BURBANK. The Pliocene and Pleistocene Deposits of Maryland.

Md. Geol. Survey, Pliocene and Pleistocene. pp. 21-137, 1906.

Contains a full description of the surficial deposits of the State with many local details.

1909.

CLARK, WM. BULLOCK and others. Report of the Conservation Commission of Maryland for 1908-1909. 204 pp., 13 pls., 13 figs.

Contains descriptions of the mineral and water resources and the agricultural soils of the State.

1918.

CLARK, WM. BULLOCK. The Geography of Maryland.

Md. Geol. Survey, Vol. X, pt. 1, 1918, 127 pp.

—————————MATHEWS, E. B., and BERRY, E. W. The Surface and Underground Waters of Maryland, including Delaware and the District of Columbia.

Md. Geol. Survey, Vol. X, pt. 2, 372 pp. 1918.

THE PHYSIOGRAPHY OF TALBOT COUNTY

BY

BENJAMIN L. MILLER

INTRODUCTORY.

Talbot County lies entirely within the Atlantic Coastal Plain and it is therefore advisable to first discuss the general characteristics of this entire physiographic province.

Boundaries and limits.—The Atlantic Coastal Plain province borders the entire eastern part of the North American continent and in essential particulars is strikingly different from the provinces on either side. The eastern limit of this province is marked by the well-defined edge of the continental shelf, which forms the top of an escarpment varying in height from 5,000 to 10,000 feet or even more. This scarp edge lies at a general depth of 450 to 500 feet below sea level, but commonly the 100-fathom line is regarded the boundary of the continental shelf. The descent from that line to the greater ocean depths is abrupt; at Cape Hatteras there is an increase in depth of 9,000 feet in 13 miles, a grade as steep as that often found along the flanks of the greater mountain systems. In striking contrast to this declivity is the comparatively flat ocean bed, stretching away to the east but with slight differences in elevation. Looked at from its base the escarpment would have the appearance along the horizon of a high mountain range with a very even sky line. Here and there notches, probably produced by the streams which once flowed across the continental shelf, would be seen, but there would be no peaks nor serrated ridges.

The western limit of the Atlantic Coastal Plain is defined by a belt of crystalline rocks consisting of greatly metamorphosed igneous and sedimentary materials, ranging in age from pre-Cambrian to Silurian. These rocks form the Piedmont Plateau province.

Most of the larger streams and many of the smaller ones as they cross the western margin of the Coastal Plain are characterized by falls or rapids, and the name "fall line" has been given to this boundary on that account. Below the fall line the streams show a marked decrease in the velocity of their currents. The position of this line near the head of navigation or near the source of water power has been a very important factor in determining the location of many of the towns and cities of the Atlantic Coast, New York, Trenton, Philadelphia, Wilmington, Baltimore, Washington, Fredericksburg, Richmond, Petersburg, Raleigh, Camden, Columbia, Augusta, Macon, and Columbus being located along it. A line drawn through these places would approximately separate the Coastal Plain from the Piedmont Plateau.

Divisions.—The Atlantic Coastal Plain province is divided by the present shore line into two parts—a submerged portion known as the continental shelf or continental platform, and a subaerial portion commonly called the Coastal Plain. In some places the division line is marked by a sea cliff of moderate height, but usually the two parts grade into each other with scarcely a perceptible change and the only mark of separation is the shore line. The areas of the respective portions have changed frequently during past geologic time, owing to the shifting of the shore line eastward or westward caused by local and general depressions or elevations of moderate extent, and even at the present time such changes are in progress. Deep channels which are probably old river valleys, the continuations of valleys of existing streams, have been traced entirely across the continental shelf, at the margin of which they have cut deep gorges. The channel opposite the mouth of Hudson River is particularly well marked and has been shown to extend almost uninterruptedly to the edge of the shelf, over 100 miles southeast of the present mouth of the river. A similar channel lies opposite the mouth of Chesapeake Bay. The combined width of the submerged and subaerial portions of the Coastal Plain province is fairly uni-

form along the entire eastern border of the continent, being approximately 250 miles. In Florida and Georgia the subaerial portion is
more than 150 miles wide, while the submerged portion is very narrow and along the eastern shore of the Florida peninsula is almost
wanting. To the north the submerged portion gradually increases
in width, while the subaerial portion becomes narrower. Except in
the region of Cape Hatteras, where the submerged belt becomes
narrower, with a corresponding widening of the subaerial belt, this
gradual change continues as far north as the southern part of
Massachusetts, beyond which the subaerial portion disappears altogether through the submergence of the entire Coastal Plain province.
Off Newfoundland the continental shelf is about 300 miles wide.

Relief.—From the fall line the Coastal Plain has a gentle slope
to the southeast, generally not exceeding 5 feet to the mile, except
in the vicinity of the Piedmont Plateau, where the slope is in places
as great as 10 to 15 feet to the mile or even more. The submerged
portion is monotonously flat, as deposition has destroyed most of the
irregularities produced by erosion when this portion formed a part
of the land area. The moderate elevation of the subaerial portion,
which in few cases reaches 400 feet and is for the most part less
than half that amount, has prevented the streams from cutting
valleys of more than moderate depth. Throughout the greater portion of the area the relief is inconsiderable, the streams flowing in
open valleys at a level only slightly lower than that of the broad,
flat divides. The country, however, shows considerable relief in
certain regions along the stream courses, though the variations in .
altitude cover only a few hundred feet.

Drainage.—The land portion of the Coastal Plain province—the
subaerial division—is marked by the presence of many bays and
estuaries representing submerged valleys of streams carved out
during a time when the belt stood at a higher level than at present.
Chesapeake Bay, which is the old valley of Susquehanna River; and
Delaware Bay, the extended valley of Delaware River, together with

such tributary streams as Patuxent, Potomac, York, and James rivers, are examples of such bays and estuaries, and there are many others of less importance. The streams which have their sources in regions to the west are almost invariably turned in a direction roughly parallel to the strike of the formations as they pass out upon the Coastal Plain. With this exception the structure of the formations and the character of the materials have had little effect on stream development except locally.

Structure.—The structure of the Coastal Plain is extremely simple, the overlapping beds having almost universally a south-easterly dip of a few feet to the mile.

Character of materials.—The materials of which the Coastal Plain is composed are boulders, pebbles, sand, clay, and marl, mostly loose, but locally indurated. In age the formations range from Cretaceous to Recent. Since the oldest formations of the province were laid down there have been many periods of deposition alternating with intervals of erosion. The sea advanced and retreated to different points in different parts of the region, so that few of the formations can now be traced by outcropping beds throughout the Coastal Plain. Differing conditions thus prevailed during each period, producing great variety in the deposits.

TOPOGRAPHIC DESCRIPTION OF TALBOT COUNTY.

The most prominent features of the topography of Talbot County are the numerous tide-water bays, creeks, and rivers that indent its shores and extend in some cases many miles inland. Ignoring them the limiting borders of Talbot County are more than 100 miles in length and of that distance tidal streams or bays extend for more than 90 miles. In other words the country is nine-tenths surrounded by tide water. If all the inlets were to be measured it would probably be found that the county has as much as 400 or 500 miles of shore line where the land is washed by tide water. Except along the Choptank River the land bordering the bodies of tide water is low and flat for many miles, consequently a great portion of the county lies only a few feet above tide. Almost one-half of the county has an elevation of less than 25 feet while the highest elevations found in the vicinity of Easton and Woodland are slightly more than 70 feet. Hills are practically absent except along the head waters of the estuaries where in a few places one can get almost the entire range of elevation in a comparatively short distance. In the lower-lying portions of the county one can scarcely observe any irregularities and even in the higher portions of the county the land is level and featureless over large areas.

TOPOGRAPHIC FEATURES.

Talbot County exhibits three general topographic features which are usually distinct. These vary greatly in the amount of the surface which they occupy and in their physical characteristics but the principal distinction is that they are found at different elevations.

Tide Marshes.

The first of these topographic features to be described consists of the tide marshes that are commonly present at the heads of the estuaries or at the mouths of tributary streams. Along the Choptank River these marshes are especially prominent and the river pursues a meandering course through them. They are sometimes present, however, even along the Bay shore. Such examples may be seen on Tilghman's Island and near Sherwood. In the region lying to the south of Talbot County there are numerous islands, some of considerable size that are completely submerged at high tide. The only island of this character represented on the map of Talbot County is a small one in Dickinson Bay on the Choptank River.

These tide marshes contain an abundant growth of sedges and other marsh plants which aid in filling up the depressions by serving as obstructions to retain the mud which the streams carry in and by furnishing a perennial accumulation of vegetable debris.

Talbot Plain.

The term plain is used in a special sense throughout this discussion to describe the flat surfaces of subaqueous origin which frequently cover extensive areas over the stream divides and whose continuations are represented in the valleys of the larger streams as terraces. The Talbot plain borders the tide marshes and extends from sea level to an altitude of about 45 feet. It is found throughout the county along the larger streams and also along the Bay shore. It is most extensively developed in the western part of the county and extends in the form of a continuous plain as far east as Easton. It also is present as a narrowed band along the Choptank River and Tuckahoe Creek as far north as Queen Anne although it is interrupted at several points where the river has cut back to the higher-lying plain.

Originally the Talbot plain everywhere sloped down gently to the water but wave erosion in exposed places has worn away most of the lowest-lying portions so that the bodies of tide-water are now almost everywhere bordered by low wave-cut cliffs from 4 to 15 feet in height. This wearing action of the waves is almost continually in operation and with only a moderate breeze the water near the shore becomes murky due to the many fine particles washed from the land and held in suspension.

Thus the shores are being continuously worn back and the transported materials dropped in the wide estuaries. The northwest winds of the winter season seem to be most effective in this destructive work and under the usual conditions the headlands exposed on that side wear away most rapidly. Except along the shores of the bays and estuaries the Talbot plain has been only slightly affected by stream action. This is due to the low elevation of the plain and also to the comparatively short period of time that has elapsed since it emerged from beneath the waters of Chesapeake Bay.

Wicomico Plain.

The Wicomico plain lies at a higher level than the Talbot, from which it is in many places separated by an escarpment varying in height from a few feet to 20 or 25 feet. This escarpment is locally wanting, so that there seems to be a gradual transition from the Talbot plain to the Wicomico. The escarpment is found, however, in so many different places, not only in this but in adjacent regions, that there is little difficulty in determining where the separation between the two plains should be made. Facing Chesapeake Bay the escarpment is well defined on the Eastern Shore throughout Talbot, Queen Anne's, and Kent counties. In Talbot County it has an almost due north and south course extending from near Wye Mills to a short distance south of Trappe. The escarpment is most pronounced a short distance southwest of Longwoods, between Longwoods and Easton, and near Hambleton. In the Choptank River valley it is less pronounced yet can be readily recognized at many points. The escarpment is an old wave-cut cliff, worn by the waves of Chesapeake Bay and the Choptank River when they were larger than at present and naturally the larger body of water would favor greater erosive action along its shores.

The Wicomico plain slopes gently to the south. South of Trappe it is little more than 30 feet above tide but rises gently to the north and near Woodland and Cordova has an elevation of slightly more than 70 feet. Elsewhere in the State the Wicomico rises to an elevation of about 90 feet where it is likewise separated from a higher-lying plain, the Sunderland, by a wave-cut escarpment.

The Wicomico plain is older than the Talbot plain and has consequently suffered more from erosion. The streams have cut deeper valleys than those in the Talbot plain and have also widened their basins to such an extent as to destroy in a great measure the continuity of its level surface. Enough remains, however, to indicate the presence of this plain and to permit its identification wherever found.

In Talbot County the Wicomico plain is widest in the northern part of the county where it covers almost the entire area between Wye River and Tuckahoe Creek and gradually narrows southward. It is seen in its best development along the Pennsylvania Railroad between Easton and Cordova.

Drainage.

The drainage of Talbot County is comparatively simple, owing to the simple structure of the formations and the location of the region adjacent to Chesapeake Bay. The land areas are with few exceptions naturally drained and there are no swamps of any importance other than the tide marshes. In some places the land has no visible drainage, as in the low land bordering the bays and estuaries, and there the water falling on the surface escapes by underground drainage. The large proportion of sand in the geological formations of the county especially favor such drainage.

TIDE-WATER ESTUARIES.—The lower courses of almost all the larger streams emptying into Chesapeake Bay have been converted into estuaries through a submergence which has permitted tidewater to pass up the former valleys of the streams. In the early development of the county these estuaries were of great value since they are navigable several miles from their mouths and thus afford means of rapid and cheap transportation of the products of the region to market. Because of them the Chesapeake Bay region was settled long before the other portions of the state. Even the advent of railroads has not rendered them valueless, and much grain and fruit is now shipped to market on steamboats and sailing vessels which pass up these estuaries.

The water in the main channel of Chesapeake Bay along Talbot County varies in depth from 60 to 120 feet. In the Choptank River as far up as Kingston Landing the water is from 10 to 50 feet in depth. On the Tred Avon River the channel to Easton Point was dredged in 1881 to 8 feet depth at mean low water. The water in

FIG. 1.—VIEW OF WYE HOUSE, A COLONIAL MANSION IN TALBOT COUNTY.

FIG. 2.—DELIVERING CORN AT CANNERY AT EASTON.

Miles River is also deep enough for large sailing vessels to pass up nearly to the head of tide. In some of the other estuaries the silt derived from the land has made shoals in so many places that they are now navigable only by light-draft vessels.

The water in the estuaries is decidedly brackish in the lower portions but almost fresh near the head of tide water in all cases where there are incoming fresh-water streams. There is seldom any distinct current in the estuaries except such as is due to the incoming and outgoing tides, and this appears to be nearly as strong when moving upstream as when moving in the opposite direction.

MINOR STREAMS.—Besides the estuaries which form so prominent a feature in this county, there are numerous minor streams which drain into them. At the head of each of the longer estuaries there is a small stream which in almost every case is much shorter than the estuary itself. Some of the estuaries, particularly those in the western portion of the county, continue as such almost to the sources of the tributary streams. Irish, Broad, Harris, Leeds, and Lloyd creeks are examples of this type.

WATER POWER.—The fall of the streams is so slight that little water power can be utilized in the county yet in a few places the minor streams have been dammed and the power used to run small mills. Miles Creek near Trappe has been dammed in two places, while Mill Creek near Wye Mills, and another smaller creek near Cordova also have small dams along their courses. There are a few other streams heading in the Wicomico plain that would like-wise furnish sufficient power to run small mills. None of the streams lying entirely or mainly in the Talbot plain have sufficient fall to warrant the construction of dams.

TOPOGRAPHIC HISTORY.

The history of the development of the topography as it exists today is not complicated and covers several different periods, during all of which the conditions must have been very similar. It is

merely the history of the development of the plains already described as occupying different levels, and of the present drainage
channels. The plains of Talbot County are all plains of plana-
tion and deposition which have been more or less modified by the
agencies of erosion. Their deposition and subsequent elevation to
the height at which they are now found indicate merely successive
periods of depression and uplift. The drainage channels have
throughout most of their courses undergone many changes; periods
of cutting have been followed by periods of filling and the present
valleys and basins are the results of these opposing forces.

SUNDERLAND STAGE.

As the Coastal Plain was depressed in the early Pleistocene the
ocean waters gradually extended up the river valleys and then over
the lower-lying portions of the stream divides, where the waves removed the older Brandywine mantle of loose materials and either
deposited the debris farther out in the ocean or dropped it in the
estuaries produced by the drowning of the lower courses of the
streams. Sea cliffs on points exposed to wave action were gradually pushed back as long as the waters continued to advance. These
now represent the escarpment separating the Sunderland from the
Brandywine. The materials which the waves gathered from the
shore, together with other materials brought in by the streams, were
spread out in the estuaries and form the Sunderland formation.
The tendency was to destroy all irregularities produced during the
post-Brandywine erosion interval. In many places undoubtedly old
stream courses were obliterated, but the channels of the larger
streams, while probably in some places entirely filled, were in the
main left lower than the surrounding regions. Thus in the uplift
following the Sunderland deposition the larger streams reoccupied
practically the same channels they had carved out in the preceding
erosion period. They at once began to clear their channels and to
widen their valleys, so that when the next submergence occurred the

streams were eroding, as before, Tertiary and Cretaceous materials. On the divides also the Sunderland was gradually undermined and worn back.

Although the Sunderland and Brandywine plains are both wanting in Talbot County at the present time it is probable that both of them did exist there at one time but have subsequently been removed. On the Western Shore of Chesapeake Bay they are well developed and in certain places form as pronounced plains as do the Wicomico and Talbot on the Eastern Shore.

WICOMICO STAGE.

When the Coastal Plain had been above water for a considerable interval a gradual submergence again occurred, so that the ocean waters encroached on the land. This submergence seems to have been about equal in amount throughout a large portion of the district, showing that the downward movement was without tilting. The sea did not advance on the land so far as during the previous submergence. The waves beat against the shore and in many places cut cliffs into the Sunderland deposits. Throughout many portions of the Coastal Plain these old sea cliffs are still preserved as escarpments, some of them 10 to 15 feet in height. Where the waves were not sufficiently strong to cut cliffs it is somewhat difficult to locate the old shore line. During this time nearly all of the Eastern Shore and a considerable part of the Western Shore were submerged. The Sunderland deposits were largely destroyed by the advancing waves and redeposited over the floor of the Wicomico sea, though those portions lying above 90 to 100 feet were for the most part preserved. Deposition of materials brought down by streams from the adjoining land also took place.

While the Wicomico submergence permitted the silting up of the drowned stream channels, yet the deposits were not thick enough to fill them entirely. Accordingly in the uplift following the Wicomico deposition the large streams again reoccupied their former channels, with perhaps only slight changes. New streams were also developed and the Wicomico plain was more or less dissected along the water courses, the divides being at the same time gradually narrowed. This erosion period was interrupted by the Talbot submergence, which carried part of the land beneath the sea and again drowned the lower courses of the streams.

TALBOT STAGE.

The Talbot deposition did not take place over so extensive an area as had that of the Wicomico. It was confined to the old valleys and to the low stream divides where the advancing waves destroyed the Wicomico deposits. The sea cliffs were pushed back as long as the waves advanced and now stand as escarpments to mark the boundaries of the Talbot sea and estuaries, forming the Talbot-Wicomico scarp line so well developed between Trappe and Longwoods as previously described.

In some places the deposits were so thick in the old stream channels that the streams in the succeeding period of elevation and erosion found it easier to excavate new courses. Generally, however, the streams once more reoccupied their former channels and renewed the corrasive work which had been interrupted by the Talbot submergence. The Talbot plain has now in many places been rendered somewhat uneven by this erosion, yet it is less irregular than the remnants of the Brandywine, Sunderland, and Wicomico plains, which have been subjected to denudation for a much longer period of time.

RECENT STAGE.

The land probably did not long remain stationary with respect to sea level before another downward movement was inaugurated. This last subsidence is probably still in progress. Before it began the Choptank, Tred Avon, Miles, and Wye rivers, instead of being estuaries, were undoubtedly streams of varying importance lying above tide and emptying into the diminished Chesapeake Bay west of their present mouths. Whether this downward movement will continue much longer or not cannot of course be determined, but there is sufficient evidence with respect to Delaware River to show that this movement has been in progress within very recent time and undoubtedly is still going on. Many square miles that had been land before this subsidence commenced are now beneath the waters of Chesapeake Bay and its estuaries and are receiving deposits of mud and sand from the adjoining land.

THE GEOLOGY OF TALBOT COUNTY

BY

BENJAMIN L. MILLER

INTRODUCTORY.

The geologic formations represented in Talbot County range in age from Miocene to Recent. Deposition has not been continuous, yet neither of the larger geologic divisions since Eocene time is entirely unrepresented. Periods of deposition over part or the whole of the region are separated by other periods of greater or less duration in which the entire region was above water and erosion was active. Aside from the Pleistocene formations the deposits are similar in many respects. With a general northeast-southwest strike and a southeasterly dip, each formation disappears by passing under the next later one. In general, also, the shore line in each successive submergence evidently lay a short distance to the southeast of its position during the previous submergence. Thus, in passing from northwest to southeast one crosses the outcrops of the successive formations in the order of their time of deposition. There are a few exceptions to this, however, that will be noted in the descriptions which follow.

GEOLOGIC FORMATION OF TALBOT COUNTY.

System	Series	Group	Formation
Quaternary	Recent		Beach sand and marsh deposits
	Pleistocene	Columbia	Talbot
			Wicomico
Tertiary	Miocene	Chesapeake	Choptank
			Calvert

TERTIARY SYSTEM

In Maryland the Tertiary includes strata of both Eocene and Miocene age both of which extend in broad bands entirely across the state. The former, although well developed in Kent, northern Queen Anne's, Anne Arundel, Prince George's, and Charles counties, does not outcrop in Talbot County. This is due to its southeasterly dip carrying it below tide in this section. It has been penetrated, however, by several artesian well borings as will be shown in the discussion of the water resources.

THE MIOCENE FORMATIONS.

The Chesapeake Group.

The Miocene deposits of the northern portion of the Atlantic Coastal Plain were at one time believed to constitute a single unit or formation to which Darton applied the term Chesapeake because of the good exposures and typical occurrences in the Chesapeake Bay region. Later, however, when additional facts concerning the fossils and stratigraphy were collected, it was found that a threefold classification would more accurately represent the lithologic and biologic character of the deposits. Shattuck therefore in 1902 proposed that the Chesapeake formation of Darton should be divided into the St. Mary's, Choptank, and Calvert formations, and that the name formerly used should be retained as a group name. The reasons for this subdivision are fully given in the general volume on the Miocene of Maryland published by this Survey.

Of the three formations named only two outcrop on the Eastern Shore. The other one, the St. Mary's, has been reached in a deep-well boring at Crisfield and should outcrop some distance to the north were it not for the deep covering of Pleistocene materials. It certainly is not present in Talbot County however. All of the formations of the Miocene are in general similar lithologically and consist of unconsolidated sands, shell marls, and sandy clay. While in most sections there is much more diatomaceous earth in the Calvert than in the other two, yet this criterion is not everywhere sufficiently diagnostic to enable one to definitely differentiate the formations. The fossils of each division possess certain individual

FIG. 1.—VIEW OF THE WYE OAK AT WYE MILLS.

characteristics and by this means the formational lines have been mainly drawn.

In the description of the individual formations we begin with the oldest one that appears at the surface which is the Calvert.

THE CALVERT FORMATION.

Name.

The formation receives its name from Calvert County where in the well-known Calvert Cliffs bordering Chesapeake Bay its typical characters are well shown. The name was proposed in 1902 * by George B. Shattuck.

Areal Distribution.

The Calvert formation extends from New Jersey in a southwesterly direction to southern Virginia where it disappears through an overlapping of later formations. It is best developed in Maryland and northern Virginia. In the latter state it presents excellent exposures in the prominent Nomini Cliffs along the Potomac River.

In Talbot County the Calvert formation is exposed only in those places where erosion has removed the overlying Pleistocene strata. Naturally these places are in the valleys of those streams that have greatest fall. and consequently the greatest cutting power and along the estuaries where wave action has been strong and the overlying cover of surficial materials of less thickness than the height of the cliff. . The best exposures are along Tuckahoe Creek where there are many bluffs 20 to 30 feet in height that are almost entirely composed of Calvert materials. Other good exposures where less thickness can be observed occur along Wye Rive, Skipton, Lloyd, Goldsboro, Peachblossom, and Trappe creeks. Along these estuaries the Calvert beds can be seen outcropping at almost every point where the wave action was strong enough to remove the talus that would tend to obscure and conceal the outcrops. Along the creeks named and some others the overlying formation is very thin, seldom more

---

* Shattuck, George B. Science, n. s., vol. xv, 1902, p. 906.

than 5 or 6 feet thick, hence all bluffs of greater height show Calvert materials at the base. West of these places there are many cliffs much higher but there the Calvert seems to have been eroded below the present level of tide water before the deposition of the Talbot sands and gravels.

Character of the Materials.

The materials constituting the Calvert formation consist of blue, drab, and yellow clay, yellow to gray sand, gray to white diatomaceous earth, and calcareous shell marl. Between these all gradations exist. The diatomaceous earth gradually passes into fine sand by the increase of arenaceous matter, or into a clay by the addition of argillaceous material. In a similar way a sand deposit with little or no clay grades over into a deposit of clay in which the presence of sand cannot be detected. Notwithstanding this variety of materials a certain sequence of deposits is commonly observed; the basal portions of the formation consist largely of diatomaceous earth, while the upper portions are composed chiefly of sand, clays, and marls. This difference in materials is much more marked on the Western Shore where the formation has been divided into the Fairhaven diatomaceous earth and the Plum Point marls. In Talbot County it seems inadvisable to subdivide the formation as impure diatomaceous earth is found at almost all horizons, though predominating in the basal portions.

The exposures of Calvert strata along the shores of Lloyd Creek and Wye River show impure diatomaceous earth or fine mealy buff to gray sand. Along Trappe Creek there are outcrops of impure diatomaceous earth or drab clay containing shell impressions and fine buff sand while a shell bed seems to be present just below tide level as fragments of shells are occasionally seen which have probably been washed up during storms. Likewise the exposures of Calvert beds along Peachblossom Creek show a drab diatomaceous clay. Along the Tuckahoe Creek there are many good exposures and the following sections are characteristic.

SECTION ON TUCKAHOE CREEK 1 MILE SOUTH OF STONY POINT.

Pleistocene. Feet

Talbot. Upper part of bank concealed by vegetation.

Gravel band composed of small quartz pebbles and numerous white clay pellets............................... ⅞

Miocene.

Calvert. Shell marl composed of rotten shells in matrix of buff sand. Numerous specimens of Venus, Melina, Pecten, Astarte, Balanus, Corbula, Crassatellites, Polynices, Turritella, etc. 3½

Fine drab to buff sand exposed to water's edge........... 8

Total 15½

SECTION ON TUCKAHOE CREEK 1½ MILES SOUTH OF STONY POINT.

Miocene. Feet.

Calvert. Drab to gray fine sand............................... 1½

Compact light-drab diatomaceous clay containing numerous shell impressions and small clear quartz pebbles; clay breaks into angular fragments when dry.......... 8

Fine buff to gray mealy sand containing few small pebbles, many shell impressions, and some fragments of shells. This evidently represents the shell layer of preceding section but most of the shells have been removed by solution. The calcareous material from the shells has cemented the sand in some places. Exposed to water's edge 4½

Total 14

SECTION ON TUCKAHOE CREEK 1 MILE EAST OF LEWISTOWN.

Pleistocene. Feet

Talbot. Loose coarse gray to buff sand containing many pebbles... 15

Miocene.

Calvert. Pure diatomaceous earth drab when wet, white when dry, containing casts of small shells and in certain layers impressions of leaf fragments. Exposed to water's edge. 12

Total 27

Paleontologic Character.

The Calvert formation of Maryland has yielded an abundance of fossils which have been described in two volumes on the Miocene

published by this Survey. In those volumes the following forms
have been described and figured:

Mammals	22	species
Reptiles	7	"
Fishes	24	"
Crab	1	
Barnacle	1	
Ostracods	31	
Cephalopod	1	
Gastropods	110	
Amphineura	1	
Scaphopods	4	
Pelecypods	119	
Brachipods	1	
Bryozoa	8	
Vermes	1	
Hydrozoa	3	
Coral	1	
Radiolaria	21	
Foraminifera	18	
Diatoms	28	

Altogether 402 species are included and the list is not by any means
complete. The diatom list is far from complete while many other
forms of mollusca have been recognized since the volumes were published. Except in the diatomaceous earth beds where diatoms and
radiolaria are very numerous, existing in countless millions, their
exceedingly tiny shells composing the mass of the material, the
pelecypods and gastropods are the most abundant fossils and numerous specimens of both of these groups are found wherever fossiliferous strata occur. The most common Calvert fossils of the
Eastern Shore are the following which have been found in many
places in Queen Anne's and Talbot counties:

Turritella plebia
Turritella aequistriata
Caecum patuxentum
Polynices duplicatus
Polynices heros
Ecphora quadricostata
Crucibulum costatum
Cadulus thallus

Arca (Scapharca) staminea
Astarte obruta
Astarte thisphila
Venus plena
Venus campechiensis
Dosinia acetabulum
Corbula idonea
Corbula inaequalis

Strike, Dip, and Thickness.

The strike of the Calvert formation is in general from northeast to southwest though it varies in some places to an almost north and south direction. In flat regions such as prevail on the Eastern Shore the line of outcrop is approximately parallel to the strike but naturally this is not the case in a region of irregular topography.

The dip of the Calvert formation is from 9 to 11 feet to the mile toward the southeast. This varies a few degrees, however, depending upon local irregularities of deposition. Since few single beds can be traced any considerable distance the exact determination of the dip is difficult.

In the region of outcrop on the Eastern Shore the thickness of the Calvert formation is about 75 feet. The full thickness does not outcrop, however, due to the overlapping strata of the Choptank formation concealing the upper beds. In a deep well at Crisfield the Calvert seems to be about 300 feet thick.

Stratigraphic Relations.

The Calvert formation rests unconformably upon the Aquia formation of the Eocene which, however, does not appear at the surface in Talbot County. Along the Chester River in Queen Anne's County and in Anne Arundel County the Calvert is seen in contact with the green sand of the Aquia. The Calvert formation is overlain by the Choptank or by Pleistocene strata belonging to the Talbot and Wicomico formations. As the Calvert dips to the southeast it passes under the Choptank but in the vicinity of its outcrop where the later Miocene deposits are absent, it has been buried beneath the

Talbot and Wicomico beds that are much younger. The only exposures to be seen now are in places where recent erosion has removed the cover of the heterogeneous materials constituting these Pleistocene strata.

Correlation.

The Calvert formation is correlated with the Petersburg horizon in Virginia but cannot be definitely correlated with any beds lying farther south. The Carolina and Gulf Miocene beds seem to be more recent, none of them apparently being equivalent to the Calvert.

Subdivisions.

The Calvert formation has been divided into two members, known as the Fairhaven diatomaceous earth and the Plum Point marls. These are fully described in the report on the Miocene of Maryland issued by this Survey. On the western side of Chesapeake Bay, particularly in Calvert County, these members can readily be distinguished but are much less distinct on the Eastern Shore and no attempt has been made to differentiate them in Talbot County.

THE CHOPTANK FORMATION.

Name.

The formation has been so named because of its exposures along the Choptank River in this county. The name was applied to these strata in 1906 * by G. B. Shattuck.

Areal Distribution.

The Choptank formation is exposed in many places along the Choptank River and its tributaries in the southern portion of the county. The best exposures are about 2 miles south of Dover Bridge where there is an excellent section that shows the typical features of the formation. Other good exposures occur at Dover Bridge, Windyhill, Kirby Wharf, and along the banks and in the vicinity of

---

* Shattuck, G. B. Science, n. s., vol. xv, 1906, p. 906.

Dividing Creek. Formerly the Choptank formation was supposed to extend farther north and nearly all the Miocene of Talbot County was referred to this member. Recent studies, however, have shown the Miocene strata of the central and northern parts of the county to belong to the Calvert formation. In its wider distribution the Choptank formation extends from New Jersey southwestward to the Potomac River where it disappears, due to an overlap of the St. Mary's formation.

Character of Materials.

The materials composing the Choptank formation are variable. They consist of fine yellow quartz sand, bluish-green sandy clay, slate-colored clay, impure diatomaceous earth, and, at some places, ledges of slightly indurated rock. In addition to these materials, abundant fossil remains are disseminated throughout the formation. The sandy phase is most characteristic of the formation, yet in this county the Choptank strata contain about as much diatomaceous clay as they do sand. The following sections illustrate the lithology of the formation in this county.

SECTION ON CHOPTANK RIVER NEAR DOVER BRIDGE.

		Feet	Inches
Pleistocene.			
Talbot.	Surficial clay loam...............................	2	
	Compact yellowish-brown sand with many darker bands of sand	3	
	Dark-gray to drab argillaceous sand..............	2	
	Pebble band; pebbles mainly about 1½ inches in diameter with some large flattened angular fragments 6 inches in diameter; pebbles contained in matrix of drab argillaceous sand..............		3-4
	Brown sandy clay..............................	1	6
Miocene.			
Choptank.	Yellowish-brown to buff fine sand considerably stained by iron...............................	5	6
	White sand		4
	Fossil band, abundant fossils in matrix of loose fine quartz sand ranging in color from yellow, buff, gray to white. Shells are mainly entire. Abundant		

species are *Macrocallista marylandica, Venus plena, V. campechiensis, Crassatellites marylandicus, Pecten madisonius, Astarte obruta, Dosinia acetabulum, Arca staminea,* while there are occasional specimens of following species to be observed: *Ecphora quadricostata, Cardium lagueatum, Turritella plebia, T. variabilis, Polynices duplicatus, P. heros, Corbula idonea,* etc. 4   6

Fine buff to gray sand containing numerous shell fragments but few perfect shells............... 2   10

Fossil layer, fossils mainly fragmentary contained in matrix of sandy clay. *Ostrea carolinensis* abundant 1

Shell fragments in matrix of ferruginous brown sand; few perfect specimens of *Ostrea carolinensis* and *Balanus concavus* 5   0

Layer of shell fragments not sharply separated from above member; composed mainly of shells of *Ostrea carolinensis* and *Balanus concavus;* layer is firmly indurated in places. Exposed to water's edge 1

Total 27   6

SECTION ON CHOPTANK RIVER ONE-FOURTH MILE SOUTHEAST
OF MOUTH OF DIVIDING CREEK.

Pleistocene.   Feet

Talbot.   Buff sandy loam grading into next member....... 1

Chocolate-colored sandy clay with pebble layer at base 2

Miocene.

Choptank.   Buff to gray fine argillaceous sand with limonite discolorations in places........................... 4

Impure diatomaceous earth varying in color from gray to light-olive green. Contains numerous impressions of small pelecypods and gastropods while one echinoid was found. Contains many vertical joints which are filled with crusts of limonite. Exposed to water's edge...................... 6

Total 13

Paleontologic Character.

The Choptank formation is extremely fossiliferous throughout the state and a great number of fossils have been described from it.

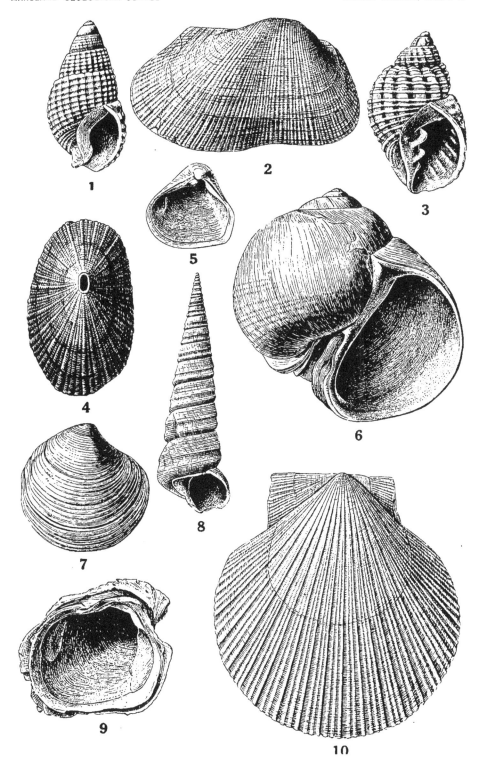

These are fully described and figured in the Miocene report of this Survey. The fauna is dominantly molluscan, pelecypods being most abundant, although there are many gastropods. In Talbot County the following species are most common:

Pecten madisonius
Macrocallista marylandica
Ostrea carolinensis
Arca staminea
Crassatellites marylandicus
Astarte obruta
Melina maxillata
Venus plena
Venus campechiensis
Cardium laqueatum
Dosinia acetabulum
Ensis ensiformis
Corbula inaequalis
Corbula idonea
Ecphora quadricostata
Polynices duplicatus
Polynices heros
Turritella variabilis
Turritella plebeia
Balanus concavus

Strike, Dip, and Thickness.

The strike of the Choptank formation is in general from northeast to southwest. The dip is toward the southeast. The amount of dip varies but in the main is about 10 feet to the mile though in some places the formation is practically horizontal. The thickness of the Choptank strata in the vicinity of outcrop is about 50 feet but it appears to thicken to the southeast after it has passed below tide as does the Calvert formation. The deep well at Crisfield seems to show about 100 feet of Choptank materials.

Stratigraphic Relations.

The Choptank formation lies unconformably upon the Calvert formation. The unconformity is in the nature of an overlap which is observable in the Calvert Cliffs of Calvert County. Above the

Choptank is the St. Mary's formation which, however, is not repre-sented in this county. In this section the formation is overlain unconformably by the Wicomico and Talbot formations which con-ceal it over the divides. The exposures occur only along the streams where erosion has removed the surficial deposits.

Correlation.

The Choptank formation cannot be correlated with any deposits lying south of the Potomac River and there is some uncertainty regarding its northward continuation in New Jersey. It seems to have a much more local development than the other Miocene forma-tions of the state.

Subdivisions.

The Choptank formation is subdivided into five zones which are distinguished from one another by the fossils they contain. These zones are most clearly differentiated in Calvert and St. Mary's counties where the formation has its greatest development. These zones, together with their fossil contents, have been fully described in the above-mentioned report on the Miocene of Maryland.

THE PLEISTOCENE FORMATIONS.

The Columbia Group.

The Pleistocene formations of the Atlantic Coastal Plain were at one time supposed to constitute a stratigraphic unit and were described under the name of Columbia formation. Later, however, it was found that the Pleistocene deposits were divisible into several formations and the name Columbia was retained as a group name while the names Sunderland, Wicomico, and Talbot were applied to the separate formations. They possess many characteristics in common, due to their origin and consist of gravels, sand, and loam.

The Columbia group in Talbot County is represented by the Wicomico and Talbot formations, the older, higher-lying Sunder-land strata having been worn away, or concealed beneath the later deposits. These form different plains or terraces, possessing very definite physiographic relations, as already described under "Topographic features."

On purely lithologic grounds it is impossible to separate the two formations composing the Columbia group in this region. The materials of each have been derived mainly from older formations in the immediate vicinity, but include more or less foreign matter brought in by streams from the Piedmont Plateau or from the Appalachian region beyond. The deposits are extremely varied, the general character changing with that of the underlying formations. Thus, materials belonging to the same formation may in different regions differ far more lithologically than the materials of two different formations lying in proximity to each other and to the common source of their material. Cartographic distinction based on lithologic differences could not fail to result in hopeless confusion. It is true that the older Pleistocene deposits are in some places more indurated and the pebbles more deccomposed than those of the younger formations; but these differences cannot be used as criteria for separating the formations, since loose and indurated, fresh and decomposed materials occur in each of them.

The fossils found in the Pleistocene deposits are far too meager to be of much service in separating the formations, even though essential differences may be shown to exist. It is the exceptional and not the normal development of the formations which has rendered the preservation of fossils possible. They consist principally of fossil plants preserved in bogs, although deposits containing great numbers of marine and estuarine mollusks have been found at a few places about Chesapeake Bay.

Physiographically the Columbia group is readily seen to consist of more than a single element. The formations occupy wave-built terraces or plains separated by wave-cut escarpments and thus indicate different periods of deposition. At the base of the escarpments the underlying Cretaceous and Tertiary formations are frequently exposed. The lowest-lying terrace is covered with Talbot materials.

In almost every place where good sections of Pleistocene materials are exposed the deposit from base to top seems to be a unit. In some places, however, certain layers or beds are sharply separated by irregular lines similar to those of a cross-bedded deposit. Some of these breaks disappear within short distances, showing clearly that they are only local phenomena in a single formation and have been produced by contemporaneous erosion or shifting shallow-water currents. Since the Pleistocene formations occupy a nearly horizontal position it would be possible to connect these separation lines if they were subaerial unconformities due to erosion, but in adjoining regions they seem to have no relation to one another. In the absence of any definite evidence showing that these lines are stratigraphic breaks separating two formations they have been disregarded. Yet it is not improbable that in Sunderland, Wicomico, and Talbot times the beds of each preceding period of deposition were in some places not entirely removed from the area covered by the advancing sea in its next transgression. Especially would materials laid down in depressions be likely to persist as isolated remnants which later were covered by the next mantle of Pleistocene deposits. If this is the case each formation from the Sunderland to the Wicomico is probably represented by fragmentary deposits beneath the later Pleistocene formations. Thus, in certain sections the lower portions may represent an earlier period of deposition than that of the overlying beds. In those regions where older materials are not exposed in the base of the escarpments each Pleistocene formation near its inner margin probably rests upon the attenuated edges of the immediately preceding formation. Since lithologic differences furnish insufficient criteria for separating these deposits and sections are not numerous enough to make a distinction to be made between local intraformational unconformities and wide spread unconformities resulting from an erosion interval, the whole mantle of Pleistocene materials at any one point is referred to one formation. The Sunderland is described as overlying the Cretaceous or Tertiary deposits and extending from the base of the Brandy-wine-Sunderland escarpment to the base of the Sunderland-Wicomico escarpment, and any possible underlying Brandywine deposits are disregarded because they are unrecognizable. Similarly the Wicomico is described as including all the gravels, sands, and clays overlying the pre-Brandywine deposits and extending from the base of the Sunderland-Wicomico escarpment to the base of the Wicomico-Talbot escarpment. Perhaps, however, materials of Talbot and Wicomico age may locally rest upon deposits of the Brandywine, Sunderland, or Wicomico formations.

THE WICOMICO FORMATION.

Name.

This formation was named from Wicomico County where it is characteristically developed.

Areal Distribution.

The Wicomico formation is co-extensive with the Wicomico Plain previously described and forms a broad band extending from the northern boundary almost to the Choptank River. With the exception of one small detached area a short distance southeast of Easton the formation is continuous and covers the main stream divide between Tuckahoe Creek and the Choptank River on the east and Chesapeake Bay and its tributary streams on the west. At one time it covered a still greater area but erosion has removed much of it and, at the present time, many streams are actively engaged in pushing their headwaters farther into the plain and removing the Wicomico materials that conceal the Tertiary formations. Many notches have been cut in this way and in several places the streams have almost succeeded in cutting through the plain, detaching portions of it. Eventually the formation will be represented by merely isolated areas of small extend lying between the streams. This already happened in many places on the western side of the Bay where erosion has been more active.

Character of the Materials.

The materials which compose the Wicomico formation consist of clay, peat, sand, gravel, and ice-borne blocks. As explained above, these, as a rule, do not lie in well-defined beds, but grade into one another both vertically and horizontally. The coarser materials, with the exception of the ice-borne boulders, have usually a cross-bedded structure, while the clays and finer materials are either developed in lenses or horizontally stratified. The erratic ice-borne blocks are scattered through the formation and may occur in the gravel beneath or the loam above. The coarser material tends to occupy the lower portions and the finer material the upper portions of the beds, but the transition from one to the other is not marked by an abrupt change and at many places coarse materials are found above in the loam and fine materials below in the gravel. The coarser materials are also frequently much decayed.

The amount of loam present in the Wicomico varies exceedingly from place to place. Wherever the loam cap is well developed the roads are firm and the land is suitable for the production of grass and grain; but where the loam is present in small quantities or absent altogether the roads are apt to be very sandy. Because of the variations in the character of the materials it is difficult to select a section that could be called typical. The following, however, is characteristic of the formation over large areas in the northern part of the county.

SECTION ALONG ROAD 1½ MILES NORTHEAST OF LONGWOODS.

Pleistocene.   Feet
Wicomico.   Yellowish-brown clay loam containing a few pebbles...... 2
            Buff to drab clay stained with limonite in places and con-
            taining many smaller clear somewhat angular quartz
            pebbles.   Exposed 1
                                                                     ———
            Total 3

Paleontologic Character.

The fossils of the Wicomico formation are limited to some plant remains and a few vertebrate bones preserved in old bogs. In Tal-

bot County, however, no fossils have yet been found in deposits of
this age.

Strike, Dip, and Thickness.

As a whole the formation occupies approximately a horizontal
position with very slight dip toward the main drainage channels.
In Talbot County the dip is toward the southwest but so small in
amount that it is difficult to determine it on account of local irregularities. The strata were laid down in many cases in old stream
valleys cut in the underlying Miocene materials and consequently
they appear to dip in the valleys and to rise in the divides. The
thickness of the formation is not at all uniform owing to the irregular surface upon which it is was deposited. It ranges from a few
feet to 50 feet or more. In Talbot County the average thickness is
from 18 to 20 feet.

Stratigraphic Relations.

In this region the Wicomico rests unconformably upon portions
of the Calvert and Choptank formations. It is in contact with the
Talbot in many places and the two formations are usually separated
by a distinct escarpment. As previously stated, in all probability
in certain places there are small portions of the Wicomico underlying the Talbot strata but it is difficult to prove that such is the
case. Wherever the Wicomico has been definitely recognized it is
exposed at the surface and has never been overlain by later deposits.

Correlation.

It represents the upper part of the later Columbia of McGee and
Darton and a part of the Pensauken of Salisbury. The presence of
glacial boulders furnishes evidence of its contemporaneity with the
ice invasion, though the particular drift sheet with which the formation should be correlated has not yet been determined.

Name.

This formation receives its name from Talbot County where it is characteristically developed. The name was first given by G. B. Shattuck in May, 1901 (Johns Hopkins Univ. Circular No. 152).

Areal Distribution.

The Talbot formation is the surface formation over the greater portion of the county. It is co-extensive with the Talbot plain already described and covers practically all the western half of the county and extends as a discontinuous terrace up the valleys of Choptank River and Tuckahoe Creek. It is dissected by numerous estuaries and in many places along the streams erosion has worn away the thin cover of Talbot materials, exposing the underlying Miocene strata. It appears as a terrace of varying width which wraps around the margin of the Wicomico formation.

Character of Materials.

The materials which compose this formation consist of clay, peat, marl, sand, gravel, and ice-borne boulders. As in the Wicomico formation these materials grade into each other both vertically and horizontally, and exhibit a tendency toward a predominance of the coarser materials in the lower part and of the finer materials near the top. There is, on the whole, a much smaller proportion of de-caped materials than in the formation just mentioned and as a result the Talbot has a much younger appearance than the Wicomico. In this county the most abundant material in the Talbot formation is drab compact clay that is stained in many places by limonite. This is especially noticeable in most of the bluffs along the shores of Chesapeake Bay and near the mouths of the larger estuaries.

Iron oxide is a common constituent of the Talbot and locally it is abundant enough to constitute a firm cementing material. The

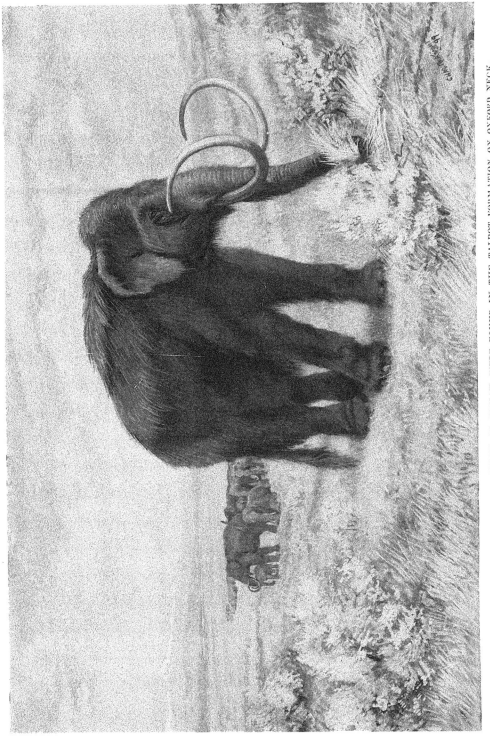

RESTORATION OF MAMMOTH, BONES AND TEETH OF WHICH WERE FOUND IN THE TALBOT FORMATION ON OXFORD NECK.

best example of this occurs near the head of Bolinbroke Creek, left bank, where there is an exposure of 18 feet of firmly indurated ferruginous sand and pebble conglomerate. Elsewhere thin bands of pebbles are cemented locally to form rocks suitable for foundations and walls. Old peat bogs are found in many places about Chesapeake Bay in the Talbot formation. In this county there are no good ones exposed yet one that shows the typical characteristics and mode of formation does outcrop near Wade's Point, 2 miles southwest of Claiborne at the base of a low bluff. The deposit is only about 6-8 inches thick and consists of twigs and stems partially lignitized.

The following sections are typical of the Talbot formation of this county:

SECTION AT TILGHMAN POINT, EASTERN BAY.

Pleistocene.   Feet

Talbot. Brown sandy loam 4

Drab argillaceous sand 3

Alternating drab, gray, and brown sand, some layers decidedly argillaceous. Exposed to water's edge............ 9

Total 16

SECTION ON CHOPTANK RIVER, 1½ MILES ABOVE KINGSTON LANDING.

Pleistocene.   Feet

Talbot. Yellowish-brown loamy sand 7

Pebble band; small quartz pebbles in matrix of drab to gray clay 4

Laminated coarse brown sand with many pebbles arranged in layers and lenses 6

Yellow sand with few pebbles......................... 1½

Compact drab clay, very hard when dry, stained with limonite in places, few pebbles...................... 5½

Pebble band, large and small pebbles in coarse gray sand ....½ to 1

Unconformity

Miocene.

Calvert. Light-buff fine loose sand greatly stained with limonite in places. Exposed to water's edge...................... 5

Total 30

SECTION ONE-EIGHTH MILE SOUTH OF BRUFFS ISLAND, MOUTH OF WYE RIVER.

Pleistocene. Feet

Talbot. Brown sandy loam containing a few pebbles.............. 1

Brown sand 2

Drab to light-brown sandy clay, very hard when dry...... 3

Loose brown to gray sand containing thin lenses of small
pebbles 4

Orange-colored sand filled with pebbles and at base many
cobbles and large boulders of quartzite, granite gneiss,
gabbro, and siliceous pebble conglomerates. Some
boulders on beach evidently derived from this layer are
4 feet in diameter. Pebbles cemented with iron in places 3

Irregular line of contact

Wicomico (?) Fine gray to greenish-gray sand containing a few small
quartz pebbles. Exposed 2

Total 15

In the above section the lower member is thought to represent a
fragment of the Wicomico that was not destroyed by the Talbot
waves but it may belong to the Talbot in which case the apparent
unconformity has been produced by contemporaneous erosion.

Paleontologic Characters.

The Talbot formation in Maryland has yielded plant remains
from the old peat bogs, marine invertebrates from a few localities
in St. Mary's, Baltimore, and Caroline counties, and vertebrate
bones from many places. So far as known no plant or invertebrate
remains of this age have ever been determined from Talbot County
although some early writers seemed to think that a deposit of oyster
shells at Easton had been deposited by natural means. It is believed,
however, that the deposit in question was made by Indians and is an
example of a kitchen midden such as occur in many places about the
shores of Chesapeake Bay and its estuaries.

Vertebrate remains have been found in this county and Cope [1]
reported mammoth, elk, deer and turtle from Talbot County. The

[1] Cope, E. D., Proc. Amer. Phil. Soc., vol. xi, 1869 (1871), pp. 171-192.

following forms from Oxford Neck are described in the Maryland Survey Report on the Pleistocene:[2]

Elephas columbi .... The southern Mammoth
Elephas primigeneus .... The northern Mammoth
Terrapene eurypygia .... Turtle
Chelydra serpentina .... Snapping Turtle

Strike, Dip, and Thickness.

The Talbot formation is similar to the Wicomico in being practically horizontal with a slight dip toward the major drainage channels. The strike accordingly changes a great deal. In Talbot County the dip is in general toward Chesapeake Bay and the strike is approximately north and south but in the southeastern portion of the county the dip is toward the south and the strike is east and west. The thickness of the Talbot varies from a few feet to 40 feet or more. The unevenness of the surface upon which it was deposited has in part caused this variability. The proximity of certain regions to the mouths of streams during the Talbot submergence also accounts for the increased thickness of the formation in such areas. In a broad band extending from north to south through the county the Talbot seems to be only a few feet thick as practically every bluff 10 feet in height along the Wye and Tred Avon rivers and Peachblossom, Trappe, and Dividing creeks shows Miocene at the base. It thickens to the westward so that no Miocene exposures occur in the western portion of the county.

Stratigraphic Relations.

The Talbot rests unconformably in different portions of the region, upon materials belonging to the Calvert and Choptank formations. As stated on a preceding page, it may rest upon deposits of Wicomico age in places although no positive evidence of this relation can be obtained. The Talbot passes beneath the waters of Chesapeake Bay and its estuaries where it is covered by Recent deposits now in progress of formation.

---

[2] Maryland Geol. Survey, Pliocene and Pleistocene. Baltimore, 1906.

Correlation.

The Talbot formation represents the lower part of the Later Columbia described by McGee and Darton and corresponds approximately to the Cape May formation of Salisbury. Its Pleistocene age is proved by the marine invertebrate fossils found in St. Mary's County and by the numerous glacial boulders scattered through the formation showing that it was formed contemporaneously with a part of the ice invasion in the northern portion of the country.

THE RECENT DEPOSITS.

In addition to the two terraces already discussed, a third one is now being formed by the currents of the rivers and the waves of the estuaries. This terrace is everywhere present along the water's edge, extending from a few feet above tide to a few feet below tide. It is the youngest and topographically the lowest of the series. Normally it lies beneath and wraps about the margin of the Talbot terrace, from which it is separated by a low scarp that as a rule does not exceed 15 to 20 feet in height. Where the Talbot formation is absent the Recent terrace may be found at the base of the Wicomico terrace. In such places, however, the scarp which separates them is higher.

Peat, clay, sand, and gravel make up the Recent deposits now forming and these materials are deposited in deltas, flood plains, beaches, bogs, dunes, bars, spits, and wave-built terraces.

INTERPRETATION OF THE GEOLOGIC RECORD.

The formations which occur within Talbot County have a more extensive development in the regions beyond its borders. Because of this fact, investigations made elsewhere in the Atlantic Coastal Plain throw much light on the conditions which have prevailed in this region during past ages.

A study of the geologic history of the county shows that it has been long and complicated. This is indicated by the many different kinds of strata represented and by the relations which they bear to one another. There are deposits that were formed in fresh or brackish waters; others that show evidence of their deposition in marine waters. Some were formed in waters of shallow depth where wave and tidal action was strong while others were laid down in quiet deep waters remote from the shore. The breaks in the conformity of the different strata also indicate that from the time of formation of the earliest beds down to the present the region has undergone many elevations and subsidences by which it has been converted into a land area or submerged beneath marine or estuarine waters.

SEDIMENTARY RECORD OF THE PRE-MIOCENE ROCKS.

In Talbot County the oldest rocks exposed at the surface belong to the Calvert formation. From deep-well borings and from observations made elsewhere in the state we know, however, that many rocks of older periods lie beneath the Miocene strata. These belong to the Eocene and Cretaceous while beneath them is the floor of crystalline rocks which appear at the surface west of a line passing through Wilmington, Baltimore, and Washington and constituting the Piedmont Plateau. The time represented by these rocks involves many millions of years during which there were many profound changes by which mountains were formed, rocks were folded, lavas were extruded, and streams cut and carved land masses as at the present day.

The crystalline rocks are so greatly folded and crushed and have been so greatly altered from their original condition that it is difficult to definitely determine their history. It is believed, however, that they represent shales, sandstones, and limestones, with some igneous rocks that have been subsequently metamorphosed to form marbles, schists, and gneisses. The rocks lying directly upon the crystalline rocks belong to the Cretaceous, or possibly Jurassic, which seems to prove that the region had remained as a land area for a very long period of time prior to the Cretaceous period or if the region were beneath the waters at any time during that interval deposits formed they were later wholly removed.

During the Cretaceous the region was elevated and depressed many times and deposits of estuarine origin and of marine origin were laid down. The Eocene strata were deposited under more uniform conditions and they consist of sands and clays in which there is a large admixture of glauconite. They were laid down in marine waters in proximity to land.

SEDIMENTARY RECORD OF THE MIOCENE FORMATIONS.

Eocene sedimentation was brought to a close by an uplift by which the shore line was carried far to the east and probably all of the present State of Maryland became land. This was followed by a resubmergence and another cycle was commenced. The deposits of the Miocene were laid down upon the land surface which had just been depressed beneath the water. Sluggish streams brought in fine sand and mud, which the waves and ocean currents spread over the sea bottom. Occasionally leaves from land plants were also carried out to sea and later dropped to the bottom as they became saturated with water.

Near the beginning of Miocene submergence, certain portions of the sea bottom received little or no materials from the land, and the water in those places was well suited as a habitat for diatoms. These must have lived in the waters in countless millions, and as they died their siliceous shells fell to the bottom and produced the beds of diatomaceous or infusorial earth which are so common in the Calvert formation. Many Protozoa as well as Mollusca lived in the same waters and their remains are plentifully distributed throughout the deposits. During the Miocene epoch the conditions seem to have been favorable for animal life, as may be inferred from the great deposits of shell marl which were then formed.

After the deposition of the Calvert formation the region was again raised and subjected to erosion for a short period, and then sank once more beneath the sea. The Choptank formation was laid down contemporaneously with the advancing ocean. This formation lies unconformably upon the Calvert, and farther north transgresses it. In neighboring regions to the south of this region a third Miocene formation, the St. Mary's, was deposited conformably upon the Choptank at a later period.

SEDIMENTARY RECORD OF THE PLEISTOCENE FORMATIONS.

At the close of the Pliocene epoch the region was raised again and extensively eroded, and was then lowered and received the deposits which constitute the first member of the Columbia group. The Sunderland, Wicomico, and Talbot formations, which make up this group, are exposed over a series of terraces lying one above another throughout the North Atlantic Coastal Plain from Raritan Bay to Potomac River, as well as in Virginia and probably still farther south.

After the close of the post-Brandywine erosion period the Coastal Plain was gradually lowered and the Sunderland sea advanced over the sinking region. The waves of this sea cut a scarp against the existing headlands of Brandywine and older rocks. This scarp was prominent in some places and obscure in others, but may be readily recognized in certain localities. As fast as the waves supplied the material, the shore and bottom currents swept it out to deeper water and deposited it so that the basal member of the Sunderland formation, a mixture of clay, sand, and gravel, represents the work of shore currents along the advancing margin of the Sunderland sea; whereas the upper member, consisting of clay and loam, was deposited by quieter currents in deeper water after the shore line had advanced some distance westward and only the finer material found its way very far out. Ice-borne boulders are also scattered through the formation at all horizons.

After the deposition of the Sunderland formation, the country was again elevated above ocean level and erosion began to wear away the Sunderland terrace. In Talbot County it was entirely removed, so far as known although, as stated on a previous page, there may be remnants of it still preserved beneath the Wicomico and Talbot sediments. This elevation, however, was not of long duration and the country eventually sank below the waves again. At this time the Wicomico sea repeated the work which had been done by the Sunderland sea except that it deposited its materials at a lower level and cut its scarp in the Sunderland formation. At this time also there was a contribution of ice-borne boulders which were deposited promiscuously over the bottom of the Wicomico sea. These are now found at many places embedded in the finer material of the Wicomico formation.

At the close of Wicomico time the country was again elevated and eroded, and then lowered to receive the deposits of the Talbot sea. The geologic activities of Talbot time were a repetition of those carried on during Sunderland and Wicomico time. The Talbot sea cut its scarp in the Wicomico formation, or in some places removed the Wicomico completely and cut into the Sunderland and still older deposits. Deposits were made on its terrace, a flat bench at the base of this escarpment. Ice-borne boulders are also extremely common in the Talbot formation, showing that blocks of ice charged with detritus from the land drifted out and deposited their load over the bottom of the Talbot sea.

Embedded in the Talbot formation at Wades Point there is a lens of drab-colored clay bearing plant remains. The stratigraphic relations of this and similar lenses of clay occuring elsewhere in the Coastal Plain show that they are invariably unconformable with the underlying formation and apparently so with the overlying sand and loams belonging to the Talbot. This relationship was very puzzling until it appeared that the apparent unconformity with the Talbot, although in a sense real, does not, however, represent an appreciable lapse of time and that, consequently, the clay lenses are actually a part of that formation. In brief, the clays carrying plant remains are regarded as lagoon deposits made in ponded stream channels and gradually buried beneath the advancing beach of the Talbot sea. The clays carrying marine and brackish-water organisms are believed to have been at first off-shore deposits made in moderately deep water, and later brackish-water deposits, formed behind a barrier beach and gradually buried by the advance of that beach toward the land.

SEDIMENTARY RECORD OF THE RECENT FORMATIONS.

The last event in the geologic history of the region was a downward movement, which is still in progress. It is this which has produced the estuaries and tide-water marshes that form conspicuous features of the existing topography. At the present time the waves of the Atlantic Ocean and Chesapeake Bay are at work wearing the land along their margins and depositing it on a subaqueous platform or terrace. This terrace is everywhere present in a more or less perfect state of development, and may be observed not only along the exposed shores, but also on passing up the estuaries to their heads. The materials which compose it are varied, depending both on the detritus directly surrendered by the land to the sea and on the currents which sweep along the shore. On an unbroken coast the material has a local character, while in the vicinity of a river mouth the terraces are composed of debris contributed from the entire river basin.

Besides building a terrace, the waves of the ocean and bay are cutting a sea cliff along their coast line, the height of the cliff depending not so much on the force of the breakers as on the relief of the land against which the waves beat. A low coast line yields a low sea cliff and a high coast line the reverse, and the one passes into the other as often and as abruptly as the topography changes, so that along the shore of Chesapeake Bay, high cliffs and low depressions occur in succession.

In addition to these features, bars, spits, and other shore formations of this character are being produced. If the present coast
line were elevated slightly, the subaqueous platform which is now
in process of building would appear as a well-defined terrace of
variable width, with a surface either flat or gently dipping toward
the water. This surface would everywhere fringe the shores of the
ocean and bay, as well as those of the estuaries. The sea cliff would
at first be sharp and easily distinguished, but with the lapse of
time the less conspicuous portions would gradually yield to the
leveling influences of erosion and might finally disappear altogether.
Erosion would also destroy, in large measure, the continuity of the
terrace, but as long as portions of it remained intact, the old surface could be reconstructed and the history of its origin determined.

THE MINERAL RESOURCES OF TALBOT COUNTY

BY

BENJAMIN L. MILLER

INTRODUCTORY.

The mineral resources of Talbot County are neither extensive nor especially valuable, yet the county contains some deposits that are of considerable economic importance, although they have not hitherto been very largely worked. Among the most important are clays, sands, gravels, shell marls, and diatomaceous earth. In addition the soils contribute much to the value of the region, which is primarily an agricultural one, and abundant supplies of water, readily obtainable in almost every portion of the county, form a further part of its mineral wealth.

THE NATURAL DEPOSITS

THE CLAYS.

Next to the soils the clay constitute the most valuable economic deposits of Talbot County. As already stated in the discussion of the stratigraphy of the region, several of the formations contain considerable quantities of clay. These argillaceous beds are rather generally distributed throughout the county, but, so far as known, have in recent years been worked only in the vicinity of Easton, St. Michaels, and Tilghman. The clays are found in each series of deposits represented in the region.

Miocene clay.—Although argillaceous beds occur very commonly in the Miocene strata of the county, they are generally too sandy to be of much economic importance. Considerable lime, derived from the numerous fossil shells which are either generally distributed throughout the sandy clay or concentrated in definite shell beds within the formations, also renders these clays of less value.

Pleistocene clays.—As already stated the Wicomico and Talbot formations are generally composed of coarse materials at the base of the deposits, with a rather persistent loam cap which marks the last stage of deposition during each particular submergence. This surficial loam, which is very similar in both formations, has been extensively used for the manufacture of brick at many places in Maryland, Virginia, and Pennsylvania. It is generally not more than 3 or 4 feet in thickness, yet, because of its position, many beds no more than 1 or 2 feet thick can be worked with profit. This loam is widely distributed throughout Talbot County and, though not quite co-extensive with the formations of which it forms a part, is present in almost every locality where the Wicomico and Talbot formations occupy flat divides that have not suffered much erosion since their deposition. The Talbot formation, especially, contains much workable clay in the broad flat river divides in the western part of the county. It has been worked at Easton almost continuously for over 40 years and has also been utilized at St. Michaels and Tilghmans Island. The clay is somewhat sandy in certain places but elsewhere is plastic and tough. In color it varies from drab to yellowish-brown. Because of the clay being a surficial deposit it is only necessary to remove the vegetation growing on the land but in other places the vegetable loam which has accumulated must also be removed before the clay can be dug.

In general the Pleistocene clays are adapted only to the manufacture of the common varieties of brick and tile, some are suitable for the manufacture of paving brick.

THE SANDS.

Inasmuch as the arenaceous phase predominates in almost every formation represented in the region, Talbot County contains an unlimited supply of sand. The sand of the Pleistocene formations is used locally for building purposes, but as it is so readily obtainable in all párts of the region no large pits have been opened. The most important are located in the vicinity of Easton and Queen Anne. Locally the Pleistocene sands are rich in ferruginous matter, which in some places cements the grains together, forming a ferruginous sandstone. Sands of this character possess a distinct value for road-making purposes, as they pack readily and make a firm road bed. When the material can be readily obtained in large quantities, good roads can be very economically constructed with it.

In some places the quartz sands of the Miocene seem to be pure enough for glass making and suggest the Miocene glass sands so extensively exploited in southern New Jersey. No attempt has ever been made to utilize these sands in this way in Maryland. Careful chemical analyses and physical tests, which have not been made, would be required to determine their usefulness in the glass industry.

THE GRAVELS.

The Wicomico and Talbot formations contain numerous beds of gravel widely distributed throughout the county. In many places they are rich in iron, which acts as a cementing agent, thus rendering them of considerable value as road metal. The most important gravel pits of this county are located near Easton and the material has been extensively used on the roads and streets of that vicinity.

THE BUILDING STONE.

Although the Coastal Plain formations of the region are composed almost entirely of unconsolidated materials, yet locally indurated beds are not uncommon. In the absence of any better stone these indurated ledges furnish considerable materials for the construction of foundations and walls. The gravel and coarse sand beds of the Pleistocene are, in many places, so firmly cemented by iron oxide as to form pebble conglomerates or sandstones. of considerable strength. The best example of such rock occurs near the head of Bolingbrooke Creek where there is a ledge of hard ferruginous sandstone about 18 feet in thickness.

THE MARLS.

The Calvert and Choptank formations are rich in deposits of shell marls, which are of value as fertilizers. From New Jersey to Georgia such deposits have been worked spasmodically since the early part of the last century, when their value was first determined, yet their importance in enriching the soil has never been generally recognized. They possess especially valuable fertilizing properties for soils deficient in lime. In some places the shells are mixed with so much sand that the lime forms only a small part of the deposit, but in others the amount of lime exceeds 90 per cent. Experiments show that better results have been obtained by the use of shell marl than by that of burned-stone lime. The marl acts both chemically and physically and has a beneficial effect on both clay and sandy soil. The shell marls have been dug in many places in this county during the past but owing to the increasing scarcity and cost of labor they are seldom dug now. They were formerly dug in the northeastern part of the county along Tuckahoe Creek, near Longwoods, Easton, Stumpton, and Royal Oak.

THE DIATOMACEOUS EARTH DEPOSITS.

As previously stated, the Calvert and Choptank formations of Talbot County contain many beds of impure diatomaceous earth. These may be of some value though they are much less pure than similar beds on the Patuxent River in Calvert County that have been worked sporadically for many years. Diatomaceous earth on account of its porosity and compactness, is used in water filters and as an absorbent in the manufacture of dynamite. It is reduced readily to a fine powder and makes an excellent base for polishing compounds, while its non-conductivity of heat makes it a valuable ingredient in packing for steam boilers and pipes and in the manufacture of safes, the latter being the principal use to which it is put. It has also been found to possess value in the manufacture of certain kinds of pottery.

THE WATER RESOURCES

The water supply of Talbot County is found in the springs and wells of the district. Many of the streams have been used at various times to furnish power for small mills, but little use has been made of them as sources of water supply. Both the private and public water supplies of the towns of the county are derived from wells. There are a few springs but they are of local value only.

SPRINGS.

The moderate elevation of the region and the small amount of dissection of the strata by streams are mainly responsible for the few springs of Talbot County. The most favorable places for them are where there are outcrops of the Miocene strata beneath a rather heavy mantle of unconsolidated Pleistocene materials. The water readily percolates through the overlying stratum and then flows along the contact between it and the less pervious Miocene materials beneath until it emerges in the valleys of streams that have cut through the overlying bed. From these springs some of the inhabitants obtain their entire supply of water which is usually of excellent quality but which are especially liable to contamination. The spring water, as also that in wells, is in places highly charged with mineral matter, particularly iron, sulphur, and salt, and some such waters have been placed on the market. The most important mineral spring of the county is a sulphur spring near St. Michaels. Others that have been reported are located near Easton, Trappe, Lloyds Landing, and Wittman. Most of them are chalybeate springs.

SHALLOW WELLS.

In Talbot County the shallow wells drawing, as they do, from the thin cover of Pleistocene sands, have the usual range in depth of from 8 to 20 feet, and the most common depths are about 10 to 15 feet. Locally driven wells have been sunk to depths of 40 and 45 feet, as at Skipton, but such wells are comparatively rare, and the majority of the inhabitants of the counties, except in the larger towns, use water from wells less than 20 feet deep. These shallow wells encounter water in beds of sand and gravel belonging to the Talbot and Wicomico formations. This shallow water is quite commonly marshy, coming probably from beds containing a large amount of vegetable matter, and in a few places the water tastes strongly of iron due to the amount of that mineral contained in the pebbles or nodules of the Pleistocene, but generally there is so little mineral matter in solution that the water is known as "soft."

In a few places near the shore the water contains considerable salt, enough to render it brackish. This is most noticeable on some of the low islands, but it has also been reported on the mainland near tide water. With increase in population these shallow wells are increasingly liable to pollution, and care should always be exercised in locating such wells where they will not be subject to contamination from outhouses.

ARTESIAN WELLS.

Attemps to procure artesian water in Talbot County have been uniformly successful, but flowing wells have been obtained only where the surface is less than 20 feet above sea level. The deepest artesian well in the county is the 1015-foot well at the Easton Water Works, and the shallowest is a 20-foot well at Copperville. This well is discussed by the author [1] who believes that "although the well was not driven below the Pleistocene deposits the water prob-

---

[1] Miller, B. L. Geologic Atlas of the U. S., U. S. Geol. Survey, Choptank Folio, No. 182, 1912, p. 8.

ably comes from a greater depth and finds its way to the base of the pipe through some deep-seated fissure in the Calvert strata." The flow of the well is very small and the water is highly charged with iron, a feature not unusual with the Calvert waters, but which has probably been acquired in this case through the passage of the water through the Pleistocene gravels.

The Miocene deposits are represented in this county by the Choptank formation which outcrops in some of the stream beds in the southern part of the county, and by the Calvert formation which is seen at the surface at a few places in the northern part along Wye River and its tributaries. The Choptank formation is not important as a water-bearing horizon, but the Calvert water has long been known and utilized. The water is soft and very good. An analysis of the water from the 135-foot wells at Easton is given on a subsequent page. The Calvert water levels are incapable of as close correlation as in the neighboring counties, due to the horizontal gradation of the materials. The Calvert wells vary in depth from 166 to 195 feet. A well at Cordova, about 40 feet above sea level, is 116 feet deep with the water within 8 feet of the surface. At Easton the Calvert long supplied the city with its water, the water being reached at 104 to 135 feet. These wells all flowed originally, but when the deep well was sunk the water in the shallower wells fell to 10 feet and would furnish no water while the 10-inch well was being pumped.

Two wells at Grubin Neck, 160 feet and 186 feet deep, draw from the Calvert horizon but do not flow, and at Windy Hill a well 180 feet deep has a small flow of water from the same stratum.

Two horizons have been recognized in the Eocene, one at the base of the Aquia and one at or below the base of the Nanjemoy. The basal Aquia water is in use at only three localities, while the Nanjemoy wells are more scattered.

Aquia.—The well at "The Anchorage," 265 feet deep, which has a small flow of hard water, and two other wells on Miles River Neck,

255 and 272 feet deep, which yield hard water but do not flow, are the shallowest wells in the county that draw from the Aquia water. At Oxford the basal Eocene water is 350 feet deep, hard, with no flows, and at Barkers Landing on the Choptank River the water is at a depth of 370 feet, is hard and rises to within 8 feet of the surface.

Nanjemoy.—The 195-foot well on Long Point reported by Darton (Bull. U. S. Geol. Survey No. 138, p. 133) and assigned by him to the Calvert level is probably Nanjemoy. More information concerning this well than that given by Darton could not be obtained.

About 2 miles south of the Long Point well a boring at Royal Oak 224 feet deep found water which was in a dark sand underneath a sand rock. The water is hard and slightly turbid, both qualities which are fairly constant in the Eocene water.

About 3 miles west of Easton a well was drilled on Dr. Nickerson's property to a depth of 297 feet in which the water in use was encountered in a bed of sand and gravel that was 48 feet thick, according to the following log supplied by the driller:

Recent.   Feet
    Soft brown clay...... .... 0 − 6
    Soft green sand.................................. 6 −117
Miocene.
    Soft green sand, shells......................... 117 −160
Eocene.
    Soft blue clay.................................. 160 −235
    Soft blue sand.................................. 235 −249
    Hard rock....................................... 249 −249½
    Green sand and gravel, Water.................... 249½−297

There are two wells at Trappe, 375 and 400 feet deep, and another one about 1 mile west, 311 feet deep, reaching hard water that stands at −14, −46, and −15 feet respectively. By the character of the water and the low head it would seem to be from the Eocene beds, but the shoal depths at which the water is reached operates somewhat against such a conclusion. It may be suggested here that the upper member of the Eocene is at this point making its first appearance on the Eastern Shore, nowhere cropping at the surface. This suggestion finds support in a slight thickness of clay and sand shown in the log of the Easton well, which has been referred to the Nanjemoy. It does not seem likely that the Calvert has as great a thickness as is demanded by the references of these wells of that horizon. In the Choptank Folio (loc. cit.) the author gives contours for the Eocene water level and states that the main stream is probably at the base of the Nanjemoy formation. This does not seem to be the position occupied by the main stream in this county since the main stream is thought to be that which supplies St. Michaels at 260 feet, Miles River Neck at 255 to 270 feet, and Oxford at 380 feet, with a distant well at Barkers Landing 370 feet deep. The wells at Royal Oak, 224 feet; west of Easton, 249 feet deep; and near Trappe, 311, 375, and 400 feet (the latter two are at an elevation of 50 to 65 feet, the 311-foot well at 20 feet, and all the others at less than 10 feet) are thought to be the base of the Nanjemoy. This is the level which becomes important across the river in Dorchester County and is about 75 feet above the lower water level of the Eocene. These are exactly the same relations as were observed in Caroline County, where at Denton the first Eocene stream is encountered at 270 feet and the second at 350 feet below sea level.

The wells drawing from Upper Cretaceous streams fall into two classes: First, the numerous wells on Tilghmans Island near Sherwood and near McDaniel around 340 to 400 feet in depth; second, the wells at Claiborne and Tunis, 440 and 486 feet deep respectively, and also the 1015-foot well at the Easton Water Works.

The wells of the first group draw from one of the best streams in the Upper Cretaceous, barring the Magothy, and the horizon is thought to be the Upper Matawan. A log of a well on Tilghman Island, 380 feet deep, is given below.*

---

* Fuller, M. L., and Sanford, Samuel. Record of Deep-Well Drilling for 1905, U. S. Geol. Survey, Bull. No. 298, pp. 233, 234.

WELL AT TILGHMAN.

Columbia.   Feet
    Hard buff clay......................................   0– 12
    Soft micaceous gray sand with a little glauconite.......   12– 18
    Soft dark brownish-gray micaceous sand..............   18– 40
Calvert.
    Gray sand containing shell fragments................   40– 50
    Hard gray sandy micaceous clay.....................   50–130
Eocene.
    Soft rock (glauconite with a little sand and gravel)
        A little water................................. 130–155
Monmouth.
    Hard black earth (dark clay with much glauconite, a
        little sand, some bits of shells).................. 155–340
    Hard dark sand (glauconite with coarse sand and bits
        of shells), some water......................... 340–360
Matawan.
    Hard dark sand (less glauconite and more grains of
        brownish quartz than preceding, some bits of
        shells), plenty of water........................ 360–380

In the original grouping of the materials the Monmouth was
omitted, but in view of the fact that the Monmouth at its outcrops
is known to be distinctly marine in origin its absence here would be
hard to explain. The above log contains several points of interest.
The hard sandy clay at 130 feet is the same as the "sand rock" encountered on Long Point at 195 feet, while the water at 155 feet is
probably the same as that found at the "Anchorage" at 265 feet.
The main water level, besides supplying the wells at Sherwood, of
depths around 365 feet, and those near McDaniels, 340 to 380 feet,
is found also in the well at Tunis Mills, 437 feet deep, where the
water was originally so alkaline that it foamed in the boilers. Another well drawing from the same level is located about 1½ miles
west, and the following materials were penetrated in the course of
drilling: Feet
    Soft yellow clay............................   0   – 15
    Soft blue clay..............................   15   – 65
    Sand rock, blue.............................   65   – 86
    Soft blue clay..............................   86   –179
    Hard blue rock............................. 179   –181
    Soft green sand............................ 181   –190

Soft blue clay	190	−250
Soft sandy black clay	250	−298
Gravel	298	−301 7/12
Soft sandy black clay	301 7/12−425	
Hard blue clay	425	−559
Green sand and gravel	559	−590

These wells all flow, although the amount is not very large, the maximum reported being 20 gallons per minute. At Easton this horizon was encountered at 570 to 600 feet, and at Oxford two wells 540 feet deep have flows of 25 gallons per minute from the same stream.

The well at Claiborne 444 feet deep was referred by Darton [*] in 1896 to the Magothy and the author, in the recent Choptank Folio, reassigned it to the Magothy along with the 540-foot well at Oxford. The Claiborne well and the 486-foot well at Tunis are here thought to be possibly Magothy, but the Oxford wells are quite certainly no lower than the upper Matawan. To correlate the Oxford wells with that at Claiborne would require that the dip be less than 10 feet in the mile, which is not at all in accordance with observation at the outcrops nor with the experience in wells. The well at Tunis is about 1½ miles down the dip from the Claiborne well and the difference in depth after correction for elevation is 40 feet. This would give a dip of about 25 feet in the mile, which is exactly the average dip of the beds at their outcrop.

The Easton well, the log of which is appended, was sunk with the object of augmenting the supply drawn from the shallow wells in the Calvert.

WELL OF THE EASTON WATER WORKS.

		Feet
Made ground	.0	− 5
Marl with whole shells	5	− 35
Light-green clay	35	− 50
Dark-brown clay	50	− 80
Dark-green clay	80	− 99
Sand rock	99	−100

[*] Darton, N. H. Bull. U. S. Geol. Survey, No. 138, 1896, p. 133.

Green sandy clay	100	−104
Light-gray sand, water-bearing, containing shark teeth, shells, large pieces of bone. Pumps 120 galls per minute and heads up to —10 feet †	104	−135
Soft light-green sandy clay	135	−170
Soft sandstone	170	−171
Soft green sandy clay	171	−190
Sandstone	190	−190½
Soft green clay, black specks	190½	−270
Hard blue clay	270	−271
Light coarse sand	271	−280
Soft rock or hard blue clay	280	−282½
Clean light-green clay	282½	−299
Boulder	299	−300½
Black sand or marl	300½	−306
Hard crust	306	−307
Black sandy clay or marl, no shells	307	−390
Black sandy clay or marl	390	−570
Sand and gravel, yellow, white, and black, water-bearing. Pumps 20 gallons per minute and heads up to —6 feet	570	−600
Fine black sand with boulders	600	−651
White sandy clay with boulders	651	−681 5/6
White clay	681 5/6	−688 5/12
Soft sandstone	688 5/12	−689
White sandy clay, hard streaks and gravel (?)	689	−727 2/3
Soft sandstone	727 2/3	−728 1/6
White clay	728 1/6	−729½
Soft sandstone	729½	−730 5/12
Soft green and brown clay with black sand	730 5/12	−802
Yellow clayey sand, similar to that at 570-600 feet	802	−835
White clay or marl, with hard and soft streaks, containing shells and sand	835	−870
Sandy clay, gray when dry	870	−888
Soft black sandy clay	888	−960
Soft rock	960	−960½
Black sandy clay	960½	−966
Soft rock	966	−967
Black sandy clay	967	−995
White sand, water-bearing, well flows	995	−1015

† During about 1885 when the Water Works were installed, six 4-inch wells were put down to this same stratum, all of which flowed at that time. When the 10-inch well was sunk it was found that the water only rose to within 10 feet of the ground, and that the smaller wells would furnish no water when the large one was being pumped.

The well passed through the Calvert water horizon, the upper Matawan level at 570 to 600 feet, and is drawing from the main water horizon of the Magothy. It should be stated here that when Darton spoke of the Potomac waters he included the Raritan which he probably thought was the horizon of this well, and it has only been recently that the Raritan has been assigned to its proper place at the base of the Upper Cretaceous. With this in mind the reference of the Easton well to the Potomac group is not so much at variance with the present assignment.

In the above log the 80-foot thickness of "green clay, black specks" may be too thick to be referred to the Nanjemoy, but the lithology is very strikingly like the clays of the Nanjemoy. There is possibly a division in this bed that was not noticed by the driller, which may also be true of the 180 feet of black sandy clay or marl at 390 to 570 feet, since some slight thickness of Monmouth probably occurs. The bed of "yellow clayey sand" at 802 to 835 feet may be the level that supplies the wells at Claiborne and Tunis Mills. The water from the white sand is plentiful and strongly alkaline. It is satisfactory for use when mixed with that from the shallow wells, but has not been tried alone.

This Magothy horizon is the largest one that has been encountered in the county thus far. The Patuxent water level is probably more than 1,000 feet below the Magothy and should yield large flows of water which, however, would probably be highly mineralized. The Magothy beds dip about 30 feet in the mile to the southeast, so that, with the Easton well for a guide, other localities can judge from its depth. After the Magothy, the next horizon in importance is the one in the upper Matawan, that which is shown in the Easton well at 570 to 600 feet, and on Tilghman's Island at a little less than 400 feet. This bed dips to the southeast at about 25 feet to the mile, so that with the Tunis Mills, Oxford, and Easton wells for guides, it too may be easily located. After these two come the Eocene and Calvert horizons which have so far never failed to yield water, although the quantity and quality have not always been satisfactory.

THE SOILS OF TALBOT COUNTY

BY

HUGH H. BENNETT, W. E. THARP, W. S. LYMAN, and
H. L. WESTOVER

INTRODUCTORY.

Talbot County lies wholly within the physiographic division known as the Atlantic Coastal Plain. The land surface varies in its topographic features from the nearly flat foreland county bordering the Chesapeake Bay and its estuaries, to the gently and moderately rolling upland plain, including most of the county east of a line drawn from Easton to Cambridge, Dorchester County. Generally there is not an abrupt break between these topographic divisions; the change is more of a gradual rise of the lower division toward the upper plain, the two blending in gentle slopes. However, there is in some places an escarpment of sufficient slope to give rise to considerable erosion. The true lower division lies largely below the 20-foot elevation and entirely below the 30-foot line. While its surface is for the most part flat, some of it is undulating. The foreland country has been indented by the streams and bays branching off from Chesapeake Bay, which have divided the country into long, narrow peninsulas and islands, making travel by land circuitous. Along the shore of the Chesapeake and its larger estuaries the waves are gradually cutting back the shore line—in some places as much as 20 feet a year—especially where the shore line is precipitous. The eroded material is carried to more protected places and deposited, eventually forming marsh.

From the very irregular shore line it might be inferred that the foreland was excessively marshy and unfit for cultivation. Such is far from true, the extent of marsh being confined to inconsiderable marginal fringes, amounting to 2.5 per cent of the land area. There are in protected places like bays and coves a few exceptionally large bodies of marsh containing several acres. Bordering the upper sources of the larger streams, like the Choptank River and Tucka-hoe Creek, there are marginal strips subject to tidal overflow vary-ing from a few rods to three-quarters of a mile in width. Very few drainage ways reach up into the interior of the foreland plain, thus leaving the flat lands without sufficient outlets for good surface drainage.

The upland plain division for the most part is gently or mod-erately rolling, and lies well for the use of modern farm machinery. In the northern part of the county elevation of 70 feet above sea level are attained at a few points. Inequalities in the surface con-figuration become less bold toward the south, changing gradually from the moderately rolling topography in the northern part to the flat and gently rolling country in the southern part, where the highest elevations are about 60 feet. There is a tendency for this upland country to assume more level and unbroken topography to-ward the interior, where is found a plain more nearly as it existed just subsequent to the emergence of the area. A larger proportion of this interior would be much more poorly drained but for the excellent underdrainage afforded by the porosity of the underlying material. Streams have not worked out good channels in this sec-tion, which fact accounts largely for the poor drainage conditions existing in the intervening flat lands.

The channels of main upland streams increase gradually from mere shallow drainage ways near the interior to comparatively deep valleys toward the marginal portion of the upland plain. The sides of some of these are sufficiently steep to have developed, through excessive erosion, a relatively broken surface. Numerous tribu-taries reach out from the main streams, affording a good drainage system to most of the upland country. The fall of many of the creeks is sufficient to develop considerable power. There are a number of water-power flour mills scattered throughout the uplands. The smaller streams are nearly all bordered by narrow strips of low, wet ground extending from mouth to source in the uplands, where considerable areas of flat, poorly drained land, sometimes semi-swampy in character, are found. The larger streams flow very sluggishly in a general northeast-southwest direction.

The Choptank River is navigable to a point near the Delaware line. The Tred Avon, Miles and Wye rivers are wide and navigable nearly to the head of tidewater, where they suddenly narrow to small streams which reach comparatively short distances into the uplands.

Talbot County was organized under the provincial government about 1661 and included the present domain of Queen Anne's, considerable portions of Caroline and Kent counties, and nearly all its present territory. Later the county was divided, and from it Talbot, Queen Anne's, Kent, and Caroline counties were formed.

Active settlement began about the middle of the seventeenth century. The settlers were mainly English, from whom the present population is largely descended. The toleration act of 1649 attracted quite a number of religious refugees, among them a considerable number of Quakers from Virginia and New England. Most of these colonists settled within sight of navigable water. Transportation and travel were mainly by water.

The farm houses and outhouses are quite substantial, and the fields are effectively fenced with wire or hedges. Churches and schools are conveniently located. The highways are excellent. Easton, the county seat, is the largest town, although there are several other important towns and shipping points in the county. From Easton, Baltimore is only forty-odd miles by water, while Philadelphia and New York, respectively, are 108 and 198 miles distant by rail, over the Pennsylvania Railroad. This railroad and various bus lines furnishes the transportation facilities by land for the entire area. Various steamboat lines are accessible from almost any part of the county and many freight-carrying sailing vessels ply between the various landings and Baltimore.

Throughout the county there are numerous canning factories, well situated near railroad stations and boat landings. These factories vary in their output from a few hundred cases of tomatoes to about 100,000 cases a season, besides the heavy packs of garden peas, corn, and pears.

AGRICULTURE.

Since its earliest settlement Talbot County has been pre-eminently agricultural in its pursuits. In the early days little more corn and wheat were produced than sufficed to supply the wants of the colonists, and in unfavorable seasons there were periods of distress from short crops.

At first tobacco was grown almost to the exclusion of other crops, and was long the medium of exchange. There were six tobacco warehouses in Talbot County in 1775, but with the beginning of the nineteenth century the tobacco acreage had enormously decreased, and warehouses were little used. The larger planters shipped their tobacco to England, but the smaller planters traded with the local representatives of English houses. Warehouse receipts representing the quantity of tobacco stored, like tobacco, passed as money.

The Revolutionary war hastened a change in agriculture by cutting off the export trade with the mother country and creating a demand for cereals to feed the Provincial army. Prior to this, however, farmers had begun to realize that their lands were being impoverished by cultivation to one crop and hence they began increasing the acreage of wheat. Clover and timothy were coming into favor and were rotated with cereals. It was about this time that the present three-field system of cropping appeared. Flax was grown for the fiber until some time before the civil war. Sweet potatoes have been produced on a small scale since the earlier days. Peaches began to be a crop of considerable importance about sixty-odd years ago. Cultivation of tomatoes for canning purposes began about 1872.

From 1859 the production of oats decreased until 1879, since when the crop has not had an important place in general agriculture. This crop too often has suffered from unfavorable weather conditions in June and July, filling out poorly.

The production of wheat has increased steadily and on a fairly profitable basis in spite of western competition. The total yield for Talbot County in 1849 was 272,963 bushels, while in 1879 it was 468,316 bushels, and in 1899, 846,340 bushels. Owing to more thorough preparation of the seed bed and to the freer use of manures there has been a marked increase in the average yields of wheat within the last twenty-five years.

Corn has also shown a steady increase. The production in 1849 for Talbot County was 621,980 bushels. By 1879 it had increased to 691,919 bushels, and in 1899 to 807,680 bushels.

The value of livestock has increased steadily. In 1899 the value of live stock was $759,581. Fertilizer expenditure varied from $110,001 in 1889, to $89,040 for 1899. The decreased expenditure means rather a decrease in the quality and price of material than any decrease in quantity. It appears that the quantity used has steadily increased since the introduction of this class of fertilizers following the civil war, and the indications are that in quantity, at least, there will be no marked decrease in the near future.

There were very few large plows in the area prior to the early seventies. Up to that time ridging for corn had been the practice. Since the introduction of the large chilled plows the practice has largely disappeared and the soils have been prepared deeper and more thoroughly. The three-field system, corn, wheat, and grass, had been the prevailing rotation up to the time of the introduction of this plow. The farmers are slowly growing less corn and wheat and practicing more diversification. More stock is being raised, more grass and forage crops grown, and more corn cut for shredding. The heavier, better drained soils are so well adapted to wheat that the crop continues to hold an important place. Farms are gradually being reduced in size, a fact which in itself points towards more intensive methods of farming.

At present the dominant system of agriculture practiced over a very large proportion of the county is general farming in connection with more or less trucking. On most farms situated within 4 or 5 miles of a cannery or boat landing, tomatoes are grown as an important crop, and often sugar corn, garden peas, and pears are grown for canning, while peaches, pears, asparagus, strawberries, and dewberries are grown for market. In some localities trucking is equally as important, or even more so, than general farming.

There are no farms devoted exclusively to dairying, although a number of farmers sell milk at the local markets or at the few creameries. A small number of the farmers are using separators, selling the cream to the creameries or shipping it and feeding the skimmed milk to hogs.

Hogs are raised to supply the home and local market demands. Poultry and sheep are considered important additions to general farming in certain localities. Most of the draft animals are raised on the farm.

Of the total acreage of land cutivated to crops in Talbot County about 60 per cent is seeded to wheat, about 25 per cent to corn, and the remainder to grass and miscellaneous vegetables, principally tomatoes. About 65 per cent of the total acreage cultivated to miscellaneous vegetables is used for tomatoes. The bulk of this crop is sold to local canneries, although a considerable quantity is shipped by boat to outside canneries. The variety most generally grown is the Stone—a medium large, uniformly ripening, prolific form, possessing the red color and fleshiness desired for canning purposes.

Considerable quantities of garden peas, sugar corn, and Kieffer pears are handled at the canneries. The Kieffer, the most successful pear grown, and a wonderful producer, is better suited to canning than marketing. Strawberries, raspberries, and dewberries of excellent quality do well on the lighter soils. The strawberries grown have a good reputation in the northern markets. The crop proves immensely profitable in years of short crops of strawberries elsewhere, and on the average a good margin of profit is realized. The Lucretia dewberry proves very successful. Wild huckleberries thrive on the light soils and considerable quantities are gathered for market. Good sweet and Irish potatoes are obtained on the light types.

For many years the peach crop was very profitable throughout the area, but owing to ravages of disease, particularly the "yellows," the industry has declined. The trees, unless diseased, make a rapid growth and produce abundantly in favorable years. Some orchards have not been injured—those near the water front seem especially resistant to disease. The crop is still of considerable importance. The Elberta appears to be the favorite variety at present, although numerous varieties are grown. Farmers should watch their orchards closely and burn root and branch, every tree as soon as the first indications of the "yellows" are noticed. Some good varieties of summer apples are grown for home use. Of the late varieties the Winesap, Ben Davis, and York Imperial have proved well suited to the soils and climate. It is believed that the Stayman Winesap, which is being successfully grown on similar soils in Delaware, would prove a profitable variety on the Sassafras loam and sandy loam. Scarlet clover is grown as a soil renovator and occasionally for seed. More cowpeas should be grown for seed, especially on the lighter Sassafras soils.

Notwithstanding that the better crop adaptations of soils are pretty clearly understood, there is too little specialization to accord with soil variation. Farmers everywhere recognize that the Elkton silt loam is a poor corn and tomato soil, though a fair soil for wheat and grass, yet all these crops are indiscriminately grown on it.

Although average yields of corn on this type are poor, in favorable seasons the crop does well, and it is with the expectation of such a season that farmers put in a considerable acreage every year instead of increasing the acreage of wheat and grass and keeping more stock.

The Sassafras loam and silt loam are admirably suited to wheat, corn, and grass, while the Sassafras sandy loam averages excellent yields of these crops. An inferior quality of wheat—small kernels— is expected on some of the poorer drained phases of these types. It is claimed, however, that the quality of that grown on the Elkton and Portsmouth soils is generally very good. Tomatoes do best on the Sassafras sandy loam, although the Sassafras loam makes good average yields, while the Sassafras silt loam makes a fairly good late crop. In growing tomatoes for canning there is no particular purpose in getting an early crop, except to head off frost. The canneries begin in August and run until the crop is canned. Strawberries and dewberries do best on the Sassafras sandy loam, Portsmouth sandy loam, and Portsmouth loam, and these types when available are generally selected for these crops. The Sassafras sand and loamy sand are particularly suited to garden peas, asparagus, turnips, early tomatoes, and Irish and sweet potatoes. Cultivated chestnuts do well on these soils.

The clovers do best on the better drained heavy soils. Red clover frequently dies out and is being replaced by crimson or scarlet clover and alsike. Timothy does well on the heavier soils, but should be grown in rotation with other crops. The well-drained, heavier Sassafras soils produce good crops of alfalfa. The subject of crop adaptation is taken up more in detail under the heads of the different soil types.

While most farmers practice some system of rotation, there are others who grow corn or wheat on the same land several years in succession. Under this treatment some of the fields have decreased in yield, but only a small proportion of the county has been subjected to such injudicious treatment.

FIG. 1.—VIEW OF ABANDONED FIELD COMING BACK TO LOBLOLLY PINE.

FIG. 2.—LOBLOLLY PINE SEEDLINGS IN OPENING IN FOREST NEAR ST. MICHAELS.

The prevailing schemes of crop succession—the old three and five field systems—are very well suited to the Sassafras loam, sandy loam, and silt loam. These systems include the following rotations: Corn, wheat, and grass for the three-field system, and corn, wheat, grass, wheat, and grass for the five-field system. The grass consists of timothy and red clover or timothy alone, and is usually cut once and then left for grazing. Tomatoes fit in well after corn and are followed by exceptionally good yields of wheat on account of the excellent physical and moisture conditions induced in the soil by the shading of the vines and the good manural properties of the vines and refuse fruit. It is estimated that wheat following tomatoes or a timothy-clover sod will yield an average of one-third more than if it follows corn. Tomatoes do not do so well after tomatoes. Rotations on all the types except, perhaps, the Portsmouth soil, should include crops of cowpeas or clover to be turned down green in conjunction with applications of 25 to 50 bushels of lime per acre once every three to six years, according to the condition of the soil and its power to retain organic matter. On account of the more thorough aeration of the lighter types, like Sassafras sand and loamy sand, the organic matter is likely to be depleted rapidly owing to rapid oxidation, and it is therefore necessary to grow frequent crops of cowpeas, but not so much lime is required as on the heavier soils. Good results are secured by liming grass in the fall preceding breaking for corn or wheat. Direct applications of lime to grass should be light (20 bushels to the acre), for the reason that too rapid decomposition of the turned-under vegetation results from large applications. In case large amounts of lime are used, the applications should be made to the broken soil so as to keep the lime from direct contact with the vegetable matter. A good many farmers are beginning to sow cowpeas or crimson clover in corn or crimson clover in tomatoes at the last cultivation, turning these under before planting wheat. The practice of turning under green crops has proved invaluable in building up the soils, both in connection with general farming and trucking.

Much trouble in getting good crops by employing the old method of seeding red clover on wheat in late winter is experienced. The young, tender plants, suddenly exposed to hot sunshine by cutting off the wheat close to the ground, seem to be unable to withstand the change and gradually die out. In seasons with plenty of moisture and no protracted hot spells succeeding harvesting, good crops are secured. Contrasting the clover yields obtained in the earlier days by sowing with wheat, and the good crops now obtained on the newly cleared peach orchards, with the crops obtained on other soils, it appears that the latter may have come into an unhealthy condition with respect to this crop. However, with an increased yield of wheat and consequent heavier growth and denser shading, the young plants are crowded nearer together and probably are less strong upon removal of the grain than formerly was the case. Some attribute the failures to toxic effects coming from continued use of acid-phosphate fertilizers. On the other hand there are many instances where good crops have been secured by seeding alone in the fall on thoroughly prepared limed ground. Many farmers claim they have no trouble in getting good crops by liming after breaking sod land, sowing wheat, topdressing with good barnyard manure, and then sowing the clover on the wheat in late winter or early spring. Scarlet clover can be grown with ease on most of the types, but the hay is not considered as good as red clover hay. Cowpeas can be grown on all soils, even those too light for crimson clover, and always improve the land, whether cut, grazed off, plowed under, or left as a winter cover crop. Alsike clover, which is rapidly coming into favor, will prove a valuable crop for this region. By growing cowpeas land too light for clover can be brought up to good condition for that crop.

Large quantities of commercial fertilizers are used. As a general rule, readily soluble, "quick-acting" fertilizers which produce an early growth and early ripening of the crop are most desirable. If nitrogen is needed, nitrate of soda is perhaps the best form in which it can be applied. It acts quickly but not through a long period, and for that reason is very desirable where short-season crops are concerned. In many cases it is found an advantage to apply the nitrate at two periods rather than all at once. It is well to make one application when the plants are set in the field and a second about the time the fruits begin to color. Fertilizers containing nitrogen in a slowly available form, such as cotton-seed meal or coarse, undecomposed stable manure, which do not stimulate an active growth until late in the season, are not desirable for this crop. Such fertilizers are too slow for a short-season crop like the tomato, which needs something to stimulate it at the very time it is trans-planted to the field. Such fertilizers also tend to stimulate late growth of vine at the expense of the maturity of the fruit. Potash and phosphoric acid are more conducive to the development of fruits than is nitrogen, except in the form of nitrate of soda.

Heavy dressings of stable manure tend to produce too much vine, and are seldom or never employed. If stable manure is used it is at a moderate rate, usually not more than one or two shovelfuls to a plant. This, if well decomposed and thoroughly incorporated with the soil, is very stimulating to the young pant and consequently very beneficial.

Any fertilizer used should be applied, in part at least, at the time the plants are transplanted to the field.*

Very little sodium nitrate is used for wheat and grass. Kainit is used frequently on the Elkton soils to prevent "frenching." About twenty years ago considerable "black residuum" was used to prevent "frenching," and it is claimed with good results. This material, composed of charred leather, undecomposed scrap iron, and traces of muriate of potash, the residuum left in the manufacture of potas-

---

* Farmers' Bulletin No. 220, p. 12.

sium prussiate, probably improved the structure of the soil and acted as an absorbent. It is said that phosphates help ripen crops and even force out a large number of "underlings" or stool stalks.

Fertilizers are applied at the average rate of about 300 pounds an acre for wheat, 200 pounds for corn, and 400 pounds for tomatoes. Heavier applications are made for crops like asparagus, garden peas, etc. Although commercial fertilizers are not as generally used for corn as for wheat, the bulk of barnyard manure is used for corn.

Experience of the most successful farmers shows that fertilizers are more lasting and beneficial when applied in conjunction with vegetable manure. Nitrogen, the most expensive ingredient of fertilizers, should be secured by growing cowpeas and clover, which crops gather atmospheric nitrogen through the action of bacteria living in the root nodules of these legumes. Alfalfa also stores up nitrogen in the soil. Too little home mixing of the fertilizer ingredients is done. Farmers generally buy from agents for future delivery.

Barnyard manure is the best fertilizer for general use on the soils of this section. Moderate amounts are made by using wheat straw as bedding material, though generally not enough to cover the land intended for corn. Considering the excellent quality of this form of manure and the ease with which heavy yields of forage crops can be produced, it seems strange that stock raising has not been carried on on a more important scale. A large extension of the stock industry undoubtedly would prove profitable. It should be the object to feed the bulk of corn and increased quantities of hay and forage crops to stock, carefully preserving the manure and returning it to the land. By establishing more co-operative creameries at convenient points throughout the county, butter making could be introduced on a profitable and permanent basis.

Farmers frequently claim that the soils are not well enough adapted to grass for profitable dairying. The yields of hay from the heavier types—from 1 to 2½ tons per acre—compare favorably with those of some of the most prosperous butter-making sections, and further, there could not be found anywhere soils better adapted to forage crops. It is not necessary to have a large acreage of pasture land where such yields of these can be secured. Silos, a comparatively small acreage of mixed grasses for pasturage, and a large acreage in cowpeas, clover, sorghum, timothy, etc., would solve the problem of feeding.

For wheat, breaking begins on stubble and sod land in late July and in August, and "corn land" is broken as soon as the crop can be removed. After breaking to a depth of 4 to 6 inches with a walking moldboard plow drawn by two or three horses, according to soil conditions, the ground is rolled with heavy iron rollers, then run over several times in opposite directions with spike-tooth and spring-tooth harrows and occasionally with a smoothing, an acme, or a disk harrow, then rolled again and seeded. It is a good idea to get the soil sufficiently pulverized for a hoe-drill to do good work, although it is not necessary to use this kind of a drill. Sometimes farmers get behind to such an extent that there is not sufficient time for giving corn land thorough preparation. Fairly good results are obtained by simply disking and seeding. Such land, however, is more inclined to run together and harden the following spring.

Most of the wheat crop is put in during the first half of October, although seeding sometimes begins as early as the middle of September and continues up to about the first of November.

Land for corn is plowed to an average depth of about 5 inches, rolled and prepared about as for wheat, then planted in checks, generally between April 20 and the middle of May. The heavier soils could be put in better condition by breaking in the fall, so as to expose the soil to the action of freezes and thaws. Especially is this beneficial for sod that has been packed by grazing or is in an unfavorable structural condition through depletion of its organic matter. However, such fall preparation for corn sometimes con-

flicts with the seeding of wheat and care of the late crops of tomatoes. Corn is cultivated comparatively deep the first two or three
times with a "buggy cultivator" or walking cultivator, then shallower with a walking cultivator. This frequent flat cultivation is
sufficient for all needs of the crop. There is a custom of cultivating
every other middle in going over a field after surface roots begin to
form. The object is to avoid retarding growth by leaving one-half
the surface roots uninterfered with until those in the cultivated
middle have time to recuperate. Corn is either cut and shocked in
the field or stripped of the lower blades and topped, leaving the ears
to be pulled. The wide corn-shock rows are sometimes seeded to
oats in late winter or early spring. Very little wheat is stacked,
thrashing being done from the shock.

Tomato land is prepared and the crop cultivated about the same
way as corn. The general plan is to set the plants close enough to
allow them to mat and completely shade the land, thus protecting
the soil and fruit from the hot sun. Fall plowing would be the
better plan, except on the Sassafras sand and loamy sand. Strawberries are cultivated in matted rows. The middles are cultivated
shallow in July or August, while weeds and grass are removed by
hoeing and by hand.

Whenever possible fall plowing should be practiced for all crops
on all soils, except the Sassafras sand and sandy loam, which would
not be particularly benefited except by turning under vegetable
matter. The depth of plowing should vary and should be generally
increased to 7 to 10 inches, care being taken not to increase the depth
more than an inch or two in one season. When more than this
amount of the under soil is turned to the surface injury is sometimes
done succeeding crops, owing to the fact that the lower soil or subsoil does not have time to weather out and get in good condition
during winter. This is especially true with the Elkton soils and
the poorer drained phases of the heavy Sassafras types. There are
instances where suddenly turning up a large quantity of the subsoil

has injured land for years. The depth of plowing should not be increased materially in the spring with the intention of growing a crop that season. Applications of lime immediately following deep plowing hasten improvement in the exposed subsoil material.

About 50 per cent of the farms of Talbot County are operated by owners, the remainder being cultivated largely by share tenants. The share tenant pays one-half the fertilizer and seed bill and receives one-half of the crops. Land is rented for one year. Landlords generally have a voice as to the acreage that shall be planted to different crops. Wheat and corn are by agreement more exclusively planted. Too often the grass area is restricted and the number of stock kept too limited for the production of a reasonable quantity of manure. A considerable number of rented farms could be managed more providently with respect to soil improvement This is sometimes neglected, owing to the lack of interest on the part of the tenant or because the landlord is too interested in immediate returns in wheat or corn. The average size of farms in 1899 was 137 acres for Talbot County. The price of land has increased considerably in the last fifteen years. Acreage valuation varies widely—according to the character of the soil, state of improvements, and locality. Land can be bought at lower figures on some of the poorer drained or deep sandy soils not within easy reach of shipping points.

THE SOIL TYPES

The superficial geology of the region is comparatively simple. The two main topographic divisions previously described as the lower foreland and the upland country comprise, respectively, the Talbot and the Wicomico plains, geologic terms applied to the younger and older series of beds of unindurated materials from which all the soils of the area have been derived. These divisions belong to the Columbia group of Pleistocene deposits, and their elevation to the present altitude above sea level is comparatively recent in a geological sense.

The chief soil-forming materials of both the Wicomico and Talbot formations are sand and silt, the latter being made up of soil grains ranging in sizes between very fine sand and clay. The former is the dominant constituent of all the soils of the northeastern part of Talbot County. Silt is the most prominent constituent in the soils of the "necks" and foreland country and of many areas, especially the flat stretches, in the uplands west of the Choptank River.

The underlying or substratum materials usually consist of sand or sand and gravel much coarser than the constituents of the overlying mass. Below a depth of about 3 or 4 feet such beds of coarse material frequently alternate or are interstratified with beds of silty clay, fine sand, coarse gravel, etc. In some sectional exposures there are exhibited alternating strata of various thicknesses—from thin seams to 2 or more feet—presenting great variety in texture and color. At Downes Landing, in Tuckahoe Neck, just outside of the county to the east, in a vertical exposure of about 15 feet, there can be seen some twenty distinct strata, which separately include silty clay, clay loam, silt loam, coarse, medium, and fine gravel, coarse, medium, and fine sands, sandy loams, fine sandy loams, and gravel and sand mixtures, covering nearly the whole range of soil classes. Cross bedding and interstratification is common in those substrata where sand is the chief constituent. The character of these lower materials does not, as a rule, affect the character of the soil, except as regards drainage conditions.

The source of most of the superficial material undoubtedly is the glaciated region and the region of crystalline rocks to the north. The sand does not have the appearance of being an old sand; it has not suffered an extreme degree of weathering. The quartz grains are subangular or much less rounded, generally, than the worn grains of some of the older Norfolk sand of the Coastal Plain region to the south, and minute mica flakes, grains of feldspar, magnetite, and fragments of dark-colored rocks are fairly abundant.

The silt deposits have a marked resemblance to some of the loess of the Mississippi Valley, particularly in structure, texture, and color. This material is of common occurrence at all elevations, but east of the Choptank River it loses its identity as a stratigraphic unit, being simply a component of the more abundant coarser soil material. There is very little silt in the deep sand deposits that occur along the east banks of the river and larger creeks, but as the distance from the streams increases the silt content of the soils increases.

Erosion, weathering, and drainage have been the most potent factors in the modification of the original material. The rate of erosion has been restricted in a large degree by the general slope of the surface and the original heavy forest growth. Erosion has been limited to a comparatively slow movement or gentle shifting of the superficial materials in the direction of the surface slope. The finer materials are moved faster than the coarser ones, and therefore a tolerably definite relationship exists between the soils and the topography. The lighter soils are found on the slopes, while the heavier soils occur in the more nearly level areas; but the transition from one type to another is always a gradual one. On broad areas of comparatively level land, where little or no surface wash has taken place, the texture has been determined by the character of the originally deposited material.

The various types are quite regular in profile, uniformity of texture, and structure, irrespective of topography or geological relationship. Generally, the surface foot carries more coarse material and is not as compact as the portion between 12 and 30 inches, and the section below 30 inches is much coarser and more open than the overlying mass. The brown color of the soil and the reddish-brown or reddish-yellow colors of the subsoil tend to give way to grayish in the soil and more nearly yellow in the subsoil toward the south.

Wherever topography and texture have combined to insure good natural surface and underdrainage, the iron content has reached a higher degree of oxidation and the soil grains have been stained brown, reddish yellow, or reddish brown. These colors in the soil and subsoil invariably indicate that condition of mineral and organic constituents which may be considered the normal state of a good, productive soil in this region. Such thoroughly aerated and oxidized soils have very few if any undesirable chemical or physical properties and are well suited to general farming. All the soils of this character have been grouped in the Sassafras series. They are the most productive and easiest managed soils of the county.

Where more or less swampy conditions have prevailed, decaying vegetable matter has accumulated, usually in sufficient quantity to form an appreciable part of the soil mass. As is common in such wet, boggy places, the accumulated vegetable matter is very black and occurs in varying stages of decomposition, from slightly changed to well decayed, mingled with earthy material, so as to make a sponge-like mass. Under natural wet conditions such soil is unsuited to most cultivated crops, but owing to the fact that the organic-matter content is otherwise in good condition, drainage only is required to bring these areas into a productive condition. Owing to the saturation of the subsoil, air has been excluded and naturally this lower material is in an unoxidized condition not very unlike that obtaining in the deeper portion of the Elkton soils. This black soil has been assigned to the Portsmouth series.

In those wet and depressed areas where the surface drainage, and generally the underdrainage, has been imperfect, the original material, subjected to intermittent wet and dry stages, has undergone unfavorable structural and chemical changes; the organic matter, though considerable in amount, is in an unfavorable condition, and the soils have turned almost white in color. Through lack of aeration the finer particles have combined rather than granulated, forming a compact, clammy mass. The absence of brown and red colors shows the iron to be in a low state of oxidation. These abnormal processes of weathering have combined to veil the properties of the original material and to bring about changes unfavorable to the development of a good agricultural soil, giving rise to the distinct series of Elkton soils. These Elkton soils stand between the Sassafras and Portsmouth soils as a transitional series that has been derived from the same material but subjected to different processes, or rather abnormal processes, of weathering.

In this grouping of the soils in series according to their most prominent characteristics of color, drainage condition, organic-matter content, productiveness, structure, etc., no account has been taken of the textural differences due to the various sizes or grades of the constituent soil grains. However, to assist in a more specific and clearer treatment, the several series have been divided into classes—sands, sandy loams, fine sandy loams, loams, silt loams, etc., according to their respective textures or relative content of coarse, medium, and fine sand, silt, and clay, as shown by mechanical separation and weighing of the various constituents of representative samples.

The following classification shows the soils of the area grouped according to processes of weathering or alteration in the original marine sediments:

Soils formed under good drainage conditions...
- Sassafras sand.
- Sassafras loamy sand.
- Sassafras sandy loam.
- Sassafras fine sandy loam.
- Sassafras loam.
- Sassafras silt loam.

Soils formed under intermittent wet and dry drainage conditions
- Elkton sandy loam.
- Elkton loam.
- Elkton silt loam.

Soils formed under swampy drainage conditions..Portsmouth sandy loam.

Unclassified alluvium and semiswampy upland...Meadow.

Alluvium subjected to tidal overflow..........Tidal marsh.

The local names of soils have been brought out in their proper relationship to the several types in so far as these names are sufficiently definite to admit proper correlation in this detailed soil classification.

The Sassafras soils are confined largely to the upland plain, although several members of the series occur in small areas in the lower foreland. The Portsmouth soil is confined almost entirely to the upland. The Elkton soils are found throughout both the low forelands and the upland country. The extent and location of the various types are shown on the accompanying map made on a scale of 1 inch to the mile. The general lay of the land is also shown on the map by contour lines drawn through points of equal elevation above sea level, and thus, besides showing the character, extent, and location of the several soils, the topographic relief of the entire country, the direction of natural drainage, and the proper location for artificial drainage ways are indicated.

SASSAFRAS SAND.

The surface soil of the Sassafras sand to a depth of 5 to 10 inches is a dull-brown sand, with a predominance of the coarse and medium grades. The subsoil is a reddish-yellow, sometimes an orange-yellow, sand which generally becomes slightly loamy and coarser toward the lower portion, frequently being underlain at about 32 inches by a reddish-yellow or reddish-brown sandy loam or sticky coarse sand. The underlying substratum is quite variable in its texture and profile features. Generally it consists of a succession of strata and seams of silty clay, coarse, medium, and fine, loose, or very compact sands or gravelly sands, and fine, medium, and coarse gravel, which vary in thickness from an inch to about 3 feet and in color from light and bluish gray to a deep reddish brown. Although in its mineralogical composition the soil material is mainly quartz, close examination reveals the presence of other minerals. Generally the finer particles cling to the larger grains in a way that tends to impart more coherence between the constituents than in case of the loose, incoherent Norfolk sand which covers extensive areas in other parts of the Coastal Plain. The Sassafras sand has not been so thoroughly reworked and washed as the latter soil and, therefore, is not so clean a sand. However, the grains have suffered considerable abrasion and are more or less rounded.

In small areas quartz gravel is interspersed throughout the soil mass, but not in sufficient quantity to change the character of the soil materially.

The Sassafras sand occurs only in limited areas in Talbot County. It is a well-drained, warm, early soil, well adapted to vegetables, especially early market-garden varieties. Excellent tomatoes, asparagus, Irish and sweet potatoes, garden peas, turnips, melons, and cucumbers can be easily grown. The yields depend largely upon the organic-matter content. There are very few soils that respond more quickly to applications of barnyard manure and the turning under of green crops, particularly legumes, such as cowpeas and crimson clover. Incorporations of such vegetable manures should be made at frequent intervals and in considerable quantities, as the soil is so thoroughly aerated and well drained that the decomposition of organic matter takes place at a comparatively rapid rate. Excellent crops of rye can be made after turning under cowpeas or crimson clover as green manures and applying moderate quantities of phosphate potash fertilizer. An application of about 35 bushels of air-slaked lime in conjunction with the turning under of heavy crops of vegetation, such as cowpeas or crimson clover, would materially assist in improving the structure of this soil by binding together the soil particles, so as to make it less open and porous. Although the type in its average condition of fertility gives rather moderate yields of the general farm crops, in years of normal rainfall very fair wheat and good corn returns can be obtained. Where the humus content has been kept up, as high as 20 bushels of wheat and 40 bushels of corn per acre have been made under conditions of fair soil treatment. Although the grasses do not do well, heavy crops of cowpeas, crimson clover, and sorghum can be made. This is a good soil for growing cowpeas for seed. Dewberries do well and strawberries fairly well.

The soil is very easily tilled and can be kept in fair condition by applying barnyard manure and turning under green legumes once every two or three years. Crops are inclined to suffer from drought in dry seasons. This type of soil can be bought for less than the heavier soils.

The following table gives the average results of mechanical analyses of samples of the Sassafras sand:

Mechanical analyses of Sassafras Sand.

Number	Description	Fine gravel	Coarse sand	Medium sand	Fine sand	Very fine sand	Silt	Clay
		Per cent	Per cent	Per cent	Per cent	Per cent	Per cent	Per cent
17903, 17923, 17925	Soil	2.5	35.9	30.5	22.8	1.0	3.9	3.3
17909, 17924, 17926	Subsoil	2.2	35.8	30.7	23.3	.6	3.8	3.7

ELKTON SILT LOAM.

The soil of the Elkton silt loam is a very light-gray to almost white silt loam. In a field in good tilth it is loose and floury, the light color and fine texture being its most marked characteristics. It contains very little medium and practically no coarse sand. The percentage of fine sand is usually low, and there is only a moderate quantity of clay. The chief constituent is silt, which forms from 60 to 80 per cent of the soil body. When wet it is yielding under foot, in some instances quite miry, but not particularly adhesive. On drying it coheres in a firm mass which may have minute cavities interspersed through it.

The organic matter content appears to be low. In the virgin soil of the woodland there is usually 2 or 3 inches of darker colored soil slightly stained by humus, but immediately below this the material is white and powdery.

In most places there is no well-defined line of contact between the soil and subsoil. The latter has about the same texture to a depth of 12 or 15 inches, below which there is an increase in the clay content. This difference in composition is most apparent between the depths of 18 and 30 inches. It is usually observable in an exposed section and is very apparent on digging or boring into the subsoil. The subsoil forms a compact stratum, very hard when dry, and, when wet, somewhat more sticky or plastic than the soil.

The subsoil at a depth of about 3 or 4 feet frequently changes to a grayish sand, medium to coarse in texture and usually saturated. This stratum varies from 1 to 2 feet in thickness and is usually underlain by a heavy bed of clay. Frequently the sandy stratum has a thin layer of soft, white, unctuous clay in it. This stratum seems to be nearly or quite impervious and is probably the cause of the saturated condition of the overlying sand.

The subsoil proper is of a grayish color somewhat darker than that of the soil. It is generally mottled with yellow and brown iron stains, especially if pockets or seams of sandy material are present. The heaviest part of the subsoil is usually a bluish-gray color with very little mottling.

This soil attains its typical development in the western part of Talbot County. The largest areas are found in the peninsulas lying between the estuaries. The central portions of most of these necks are nearly level, and the sluggish surface drainage, together with the character of the material, accounts for the formation of this type. Smaller areas are found in the uplands wherever similar conditions prevail.

The type is associated with the Sassafras silt loam and is derived from the same kind of material. The differentiation is due entirely to drainage conditions. While the two soils are very distinct in color, organic matter content, and general agricultural value, the change from one to the other is frequently so gradual that it is difficult to draw an exact boundary.

The original vegetation consisted chiefly of white oak, and the land is now locally termed "white oak land." A mixed forest now occupies uncleared fields, and loblolly pine, which came in when the original white oaks were removed, is a common species and attains good size. Other kinds of oak trees, with gum, soft maple, beech, and dogwood, are very commonly found on this type.

Much of this soil is under cultivation. It is somewhat difficult to manage, especially in wet seasons. After a rain the soil tends to run together, and on drying forms a smooth, hard surface. When slightly moist it yields very readily to tillage.

The Elkton silt loam is well adapted to timothy and excellent crops are grown. The acreage planted to this crop is rather limited, and could be increased with profit. Red clover does not grow on this soil well, probably on account of a somewhat acid condition. In fields which have reasonably good surface drainage scarlet clover is successfully grown. A large proportion of the tillable area of this type is annually sown to wheat. The same cultural methods are generally practiced as on the Sassafras silt loam. In seasons which are not excessively wet the average yields compare well with those of the Sassafras soil. In some seasons a good crop of corn is secured, but this land is too cold, wet, and deficient in humus to be well adapted to this cereal.

The improvement of many of the small areas in the upland country requires ditches of sufficient capacity to remove promptly the excess water. Ditches should be located near the upper margin, so that the water will be intercepted and not saturate the lower lying soil, as is the case where the main ditch is placed near the center of the area. It is highly essential that the ditches be deep enough to lower the water tables to 2½ or more feet below the surface, so that the subsoil may be aerated.

The permanent improvement of the larger areas, particularly those on the "necks." is a more difficult problem. They are larger and so flat that it is sometimes expensive to construct ditches with an adequate outlet.

The nature of the strata underlying the subsoil has already been briefly described. It seems probable that the structure of the deeper subsoil is the same for much of the Elkton soil of the "necks" and for the Sassafras silt loam adjoining these areas. The roadside cuts on the slopes leading down from the upland often show a light-colored clay with sandy material between the subsoil and the deeper subsoil. Any excess of water in the soil of the higher ground tends to follow the sandy stratum to the lower levels. Where no natural drainage line intervenes between the higher ground and that somewhat lower a positive upward pressure of the ground water may exist in the latter. This is probably why the subsoil of the Elkton silt loam remains saturated long after surface conditions would indicate a normal moisture content.

The present artificial drainage consists of surface ditches. They are usually shallow—only a foot or two in depth. These serve to remove the excess of rainfall, but fail effectually to drain or prevent a waterlogged condition of the subsoil. It is highly desirable that the lower soil be so drained as to have comparatively free access of air.

It cannot be positively stated, however, that even thorough drainage alone will result in an immediate improvement of this soil. It has been so long subject to intermitten saturation that its constituents have undergone important changes. The soil has quite lost the property of granulation as is evidenced in its lifeless, puttylike feel when moist and its firm cementation when dry.

The frequent unsatisfactory crop yields from well-drained land which has been given good culture indicates a poor structural condition or the presence of injurious substances. There are very few iron concretions. The ferruginous material is seldom further concentrated than in small, soft grains and thin streaks of limonite, or other compounds of iron, and these are not abundant in the heaviest phases of this soil where aeration is least active. It is probable that the decaying organic matter under the moisture condition which usually prevails renders soluble much of the iron forming ferrous carbonate, a substance injurious to cultivated plants. Under similar conditions the phosphoric acid of the soil combines with the iron in an insoluble form, thereby becoming unavailable for crops.

Experience has shown that a heavy application of coarse manure often gives surprisingly good results. There should be an abundance of organic matter incorporated with the soil, which can be most cheaply done by plowing under some green crop. This should be done in early fall or precede by some months the time of planting the next crop, so that partial decay may take place. Otherwise more harm than good may be done the first crop. Complaint is made sometimes that saturating rains following deep fall plowing cause the soil to run together in such a way as to give rise to the formation of an almost glasslike smooth surface, which bakes and hardens with subsequent sunshiny weather. This trouble would be lessened by the incorporation of coarse manures, the application of lime, and general improvement of the drainage condition.

Lime should be liberally applied (30 to 40 bushels an acre), not only for the favorable chemical effect it has on this soil, but that flocculation of the clay may be favored. Turning under green vegetation should be done always in conjunction with an application of from 25 to 50 bushels of lime per acre.

After the physical condition of the soil has been improved in the manner thus outlined, its further fertilizer requirements depend upon the crop to be grown or the method of farming practiced. It will probably be found that little or only moderate amounts of commercial fertilizer are needed. In case fertilizer is found to be necessary, one containing a high percentage of phosphoric acid would be of benefit, especially to wheat.

Where it is impracticable to drain land of this kind it seems that such crops as timothy and redtop offer the best means of utilization.

The following table gives the average results of mechanical analyses of samples of the soil and subsoil of this type:

Mechanical analyses of Elkton silt loam.

Number	Description	Fine gravel	Coarse sand	Medium sand	Fine sand	Very fine sand	Silt	Clay
	*	*Per cent*	*Per cent*	*Per cent*	*Per cent*	*Per cent*	*Per cent*	*Per cent*
16990, 17900.......	Soil	0.2	2 3	2.5	4.8	2.6	77.5	10.0
16991, 17901.......	Subsoil	.2	3.2	3.2	5.1	2.9	63.7	23.0

A determination of the organic matter gave the following percentages: No. 16990, 3.55 per cent; No. 17900, 1.42 per cent.

SASSAFRAS SILT LOAM.

The Sassafras silt loam to a depth of 8 or 10 inches is a friable silt loam. A dry sample rubbed between the fingers breaks into a soft, pulverulent mass in which little or no medium sand can be detected. A perceptible quantity of fine sand is present, consisting in part of minute mica flakes. When wet such a sample is somewhat plastic, mulches easily, and shows little tendency to adhere. On drying it becomes crumbly, the fragments being weak and porous.

The soil yields readily to tillage, and the cultivated land has a soft, loamy surface. A considerable portion of the first 2 or 3 inches will be almost pulverulent if the field has been harrowed or rolled when in a slightly moist condition. If clods form at all, they are generally small, porous, and break under a light pressure.

The color of the moist soil is usually a yellow brown. It becomes lighter as the moisture content decreases and not infrequently approaches a buff or very light yellowish brown. An exposed section in a roadside cut generally shows, beneath an inch or two of grayish surface loam, a light-yellowish soil grading downward to a reddish yellow or dull reddish brown, which is usually the color of the subsoil.

The subsoil contains more clay than the soil and is usually rather compact. Between the depth of 15 and 30 inches it is somewhat granular. On drying it breaks into roughly angular fragments. The granulation is not strongly developed and is easily destroyed by manipulation when the material is moist.

This soil is of common occurrence in Talbot County. It ranges in altitude from 10 to 70 feet above sea level and attains its typical development on the gently undulating interstream divides. On the "necks" this type is confined to those portions which have relatively good drainage. Where the surface has little or no relief the Sassafras silt loam gradually passes into the Elkton silt loam. Wherever the surface is more rolling some of the lighter soils are generally found.

The drainage is usually good. The sandy substratum gives excellent underdrainage and prevents any undue accumulation of water where the surface is slightly depressed.

The color of the subsoil is a reliable indication of the average moisture conditions. Where the color is brown or approaches a reddish yellow the drainage is effective and the soil mass has good aeration. Where the drainage is not as thorough as it should be or is somewhat sluggish the soil usually is of a pale-yellow shade.

In some of the depressions which occur in this type the soils have a texture, structure, and color so different from that of the Sassafras series that they belong to the Elkton series. The areas are usually too small or ill defined to be shown on a map of the scale used.

All of this type was originally forested. It seems to be a congenial soil for numerous species of trees and shrubs. Most of it is now under cultivation, but in the forested portions almost every variety of tree common to this section of country may be found, excepting those confined to marshy soils.

The Sassafras silt loam is well adapted to grass, forage crops, and wheat. The average yields of grain are quite as high as upon any other type in the area. Its texture admits of the preparation of an ideal seed bed, and the average moisture content is favorable for winter wheat. Its structure admits of only a minimum loss through leaching of the fertilizer applied. Twenty-five to 30 bushels of wheat per acre is not an uncommon yield in favorable seasons, but the average is nearer 18 or 20 bushels.

Clover and timothy do well. It is sometimes difficult to secure a good stand of clover, but this trouble is due to cultural practices or seasonal extremes rather than any condition peculiar to the type.

The yields of corn in favorable seasons average from 50 to 60 bushels per acre. The yield could be improved by liberal applications of organic matter. The usual supply of barnyard manure and the frequent changes to grass fail to give the needed quantity of humus. The fact is frequently overlooked that a high organic matter content, besides assisting in the maintenance of moisture, also improves the physical condition of the soil. In this instance it would tend to prevent the "running together" of the surface soil after each heavy rain.

This is the heaviest well-drained soil in the county. It should be restricted to those crops which require a long growing season and for which a continuous moisture supply is of first importance.

The following table gives the average results of mechanical analyses of typical samples of the soil and subsoil of this type:

Mechanical analyses of Sassafras silt loam.

Number	Description	Fine gravel	Coarse sand	Medium sand	Fine sand	Very fine sand	Silt	Clay
		Per cent	*Per cent*	*Per cent*	*Per cent*	*Per cent*	*Per cent*	*Per cent*
16988, 17011, 17939	Soil	0.6	2.8	2.7	4.3	8.9	71.9	8.6
16989, 17012, 17940	Subsoil	Tr.	1.1	1.5	2.9	9.7	65.1	19.4
17013	Lower subsoil .	3.6	18.3	12.4	8.3	0.5	45.8	11.6

A determination of the organic matter gave the following percentages: No. 16988, 1.02 per cent; No. 17011, 1.07 per cent; No. 17939, 0.92 per cent.

SASSAFRAS LOAMY SAND.

The Sassafras loamy sand represents a transition between the sand and sandy loam of the Sassafras series. In agricultural value it is much inferior to the normal sandy loam of the series, but is much more productive than the sand. It presents variations which are apparent even upon casual examination, and which are of considerable importance from an agricultural standpoint. These are determined for the most part by the distribution, or location, in the soil profile of the silt and clay. In the greater proportion of this soil the fine material is found largely between the depths of 15 and 30 inches. The most extensive development of this phase occurs around Bruceville.

The soil to a depth of 6 to 8 inches is a dull-brown loamy sand. All grades of sand are found, but medium to coarse grains usually form a considerable part of the whole and give a coarse, gritty character to the material. There is some small quartz gravel, which with the larger sand grains is quite conspicuous on the surface after a rain. The proportion of fine material is usually sufficient to cause the sand to cohere feebly if a moist sample is pressed in the hand. When dry it is quite loose, but no so incoherent as the Sassafras sand.

The upper part of the subsoil has about the same texture and structure as the soil. It is much lighter in color, usually a pale yellow. At a depth of 15 inches there is a perceptible increase in the percentage of fine material. This lower part of the subsoil is a moderately heavy sandy loam. If moist it is somewhat sticky; when dry the fine material binds the sand grains in a rather friable mass. Unless exceptionally coarse it possesses good capillarity. This part of the soil section is essentially the moisture reservoir of the soil.

The surface of most of this phase is gently undulating. Some areas of considerable size are nearly level or have a low but very uniform slope toward the nearest stream. Slight inequalities of the surface frequently indicate differences in the agricultural value of the land, the high ground usually having the heavier subsoil. Not infrequently small elevations are much lighter in texture, but in general this type passes into the Sassafras sandy loam along the higher contour lines.

In some places the surface assumes a grayish tint when dry, while the subsoil approaches pale yellow in color. This indicates poor drainage. This condition is not always due to topographic position, for some of this phase is underlain by a thin clay stratum similar to that found under the Elkton silt loam. It frequently occurs at a depth of 40 or 50 inches, but seems to be local in its development.

Where drainage is necessary the proper location of ditches is somewhat difficult to determine. The depth to the clay substratum and the direction of its slope should be taken into consideration. In most instances the ditch should be dug along the upper side of the tract to be drained instead of being located in the middle or lower part.

The native timber growth comprises most of the oaks common to this area and sweet gum, dogwood, and pine, with a few birch, beech, and hickory. Alder and huckleberry are abundant undergrowths. Crabgrass commonly takes possession of neglected fields, followed later by loblolly pine.

The Sassafras loamy sand is easily cultivated and responds well to fertilization. Since all of it is very deficient in humus, the organic matter content should be increased. It would also be of benefit to have some cover crop during the winter. This prevents to some extent the excessive leaching to which these light soils are subject. An acreage application of about 30 bushels of lime in conjunction with a green crop, preferably cowpeas or clover, would improve the structure; that is, bind the soil particles into such an arrangement as would make the soil less open and droughty.

Much of this soil is now used for general farm crops. The yields of corn and wheat are low and often severely affected by dry weather. It is better adapted to truck crops and an increasing acreage is being planted. Melons, cantaloupes, tomatoes, and sweet potatoes are successfully grown. Buckwheat, crimson or scarlet clover, and cowpeas do well and many small fields of the crops are grown. Peach orchards might well be located on the well-drained portions of the soil.

The following table gives the average results of mechanical analyses of the Sassafras loamy sand:

Mechanical analyses of Sassafras loamy sand.

Number	Description	Fine gravel	Coarse sand	Medium sand	Fine sand	Very fine sand	Silt	Clay
		Per cent	*Per cent*	*Per cent*	*Per cent*	*Per cent*	*Per cent*	*Per cent*
17048, 17902.......	Soil	1.6	20.5	22.4	28.2	1.7	18.3	5.9
17049, 17903.......	Subsoil	1.5	22.3	21.8	30.2	2.6	15.3	5.7

SASSAFRAS SANDY LOAM.

The Sassafras sandy loam in its typical development to a depth from 9 to 13 inches is a grayish-brown or brown moderately heavy sandy loam, consisting of a fairly even distribution of the coarse, medium, and fine grades of sand with a relatively high proportion of silt which coheres to the sand grains so as to impart a distinctly loamy character to the soil, especially when dry. The soil always has a more pronounced sandy feel when wet, owing to a weakening of the binding power of the finer material which is given freer movement by the excess of moisture. There are some areas very much lighter than the general average as described above. The absence of very fine sand is everywhere noticeable.

The subsoil consists of a reddish-yellow or reddish-brown sandy loam or heavy sandy loam which at 26 to 30 inches generally passes into a reddish-brown coarse light sandy loam to sticky coarse sand, with small quartz gravel sometimes quite compact. The tendency is toward a slightly lighter and coarser textured subsoil, more compact and more nearly red with increasing depth. Sometimes the upper portion of the subsoil is pale yellow and siltier than the average, while in the more nearly level and poorer drained areas the pale yellow may extend as far downward as the change in the substratum. As a rule the subsoil is only slightly heavier than the surface soil, and like it carries considerable silt and little very fine sand.

Generally the well-drained soil becomes lighter in color as the organic-matter content diminishes, but fields often are spotted with gray in slight depressions, where the soil approaches the Elkton sandy loam, and may contain considerably more organic matter than the surrounding better drained and more productive soil. The better drained brown phase of the northern part of the county tends to give way to a lighter colored and somewhat less productive phase accompanying a moderation in the surface relief toward the south. Occasionally small areas are quite gravelly, especially on stream slopes. A more nearly red subsoil is found in the better drained areas. The type is locally styled "red clay bottom" or "medium light loam."

On account of the coarser substratum excellent underdrainage obtains throughout the larger proportion of the type, which feature, coupled with the good texture of the overlying material, makes the type a thoroughly aerated soil, capable of maintaining a supply of moisture favorable to healthy plant development.

Under fair treatment the soil is tilled easily under widely different moisture conditions, yielding most readily to treatment of all the extensive general purpose soils. However, continuous cultivation without restoring vegetable matter, especially where closely grazed through all sorts of weather, is apt to induce a compact structure resistant to plowing and favorable to excessive loss of moisture in dry spells by surface evaporation.

The Sassafras sandy loam occurs throughout the county and is largely confined to the uplands. It is conspicuously absent from the lower forelands. It sometimes follows the slopes of the inland stream valleys nearly to tide level, thus occupying all variations in land surface from broken stream slopes to moderately rolling and nearly flat upland country. In some sections the surface configuration is interrupted by small, poorly drained depressions, some of which contain bodies of Portsmouth or Elkton soils too small to be outlined on the map.

The type has been derived from sediments of marine deposition brought down from the region of crystalline rocks to the north of Maryland and Delaware. The sand particles are generally less round than those of the Sassafras sand, showing that the material has been subjected in a less degree to reworking by water and wind. Since the elevation of the sedimentary material above the sea considerable weathering has taken place to a depth of 3 feet or more. This weathering mainly consists in the oxidation and consequent change in color of the material and the accumulation of organic matter in the surface. There has been considerably less washing out of the fine materials and leaching than in the Norfolk sandy loam, a much less productive soil occurring extensively in other parts of the Coastal Plain.

The Sassafras sandy loam is adapted to a wider range of crops than any other type of the area. The principal crops are corn, wheat, tomatoes, and grass. Under favorable conditions of weather and soil management, and accordingly as the crop is grown on the lighter or the heavier phase, wheat yields from 15 to 30 bushels, corn from 35 to 65 bushels, hay from 1 to 2 tons, and tomatoes from 4 to 12 tons per acre.

This is the best tomato and strawberry soil of the county. Excellent returns were secured from a large acreage of tomatoes put out in 1907, averaging 5 tons per acre, and a considerable acreage of strawberries was also very profitable.

Irish potatoes, sweet potatoes, cantaloupes, watermelons, cucumbers, and asparagus do well, as do all kinds of forage crops suited to this climate. Pears, peaches, chestnuts, dewberries, raspberries, and blackberries find the soil well suited to their needs. In view of the ease with which good crops of cowpeas and crimson clover can be secured, and the readiness with which the soil responds to applications of barnyard manure, these legumes should be more generally grown in connection with an extension of the live-stock industry. Some claim that the soil is not well enough adapted to grass to yield satisfactory returns in stock raising, but by growing forage crops, excellent yields of which can be obtained, any such deficiency can be offset. A ton or two per acre of hay, however, is not considered poor even in the most prosperous stock raising sections.

Turning under green crops of cowpeas or clover grown in rotation with wheat, corn, tomatoes, or grass, or any other rotation, in conjunction with applications of 25 to 40 bushels of lime per acre every four or five years, is the most economical method of improving and maintaining the productivity of the type. Two excellent rotations are corn with cowpeas or crimson clover, wheat, grass, wheat, cowpeas, or clover to be turned under and limed, then corn; and corn, with cowpeas or crimson clover, tomatoes, wheat, grass and clover, corn. The same fertilizers are used upon this soil as upon the other soils of the area.

The following table gives the average results of mechanical analyses of the soil and subsoil and the results of a single determination of the lower subsoil of the Sassafras sandy loam:

Mechanical analyses of Sassafras sandy loam.

Number	Description	Fine gravel	Coarse sand	Medium sand	Fine sand	Very fine sand	Silt	Clay
		Per cent	Per cent	Per cent	Per cent	Per cent	Per cent	Per cent
17014, 17917.......	Soil	2.9	20.9	15.4	17.7	2.7	35.2	5.1
17015, 17918.......	Subsoil	2.9	18.5	11.9	14.1	3.0	35.7	13.9
17919	Lower subsoil .	1.3	19.2	19.3	26.1	3.9	17.5	12.2

A determination of the organic matter gave the following percentage: No. 17917, 1.05 per cent.

SASSAFRAS FINE SANDY LOAM.

The surface soil of the Sassafras fine sandy loam is a grayish-brown to yellowish-brown, quite silty fine sandy loam which grades into a pale-yellow material of the same texture a few inches beneath the surface. At about 20 inches the soil portion is underlain by a reddish-yellow, compact, rather clammy light silty loam, which in turn is underlain at 28 to 36 inches by an orange-yellow, reddish-yellow or reddish-brown, light, fine to medium sandy loam. Wherever the relief is sufficient to insure good drainage the subsoil is quite friable, admitting of good aeration and circulation of moisture. On the other hand, the subsoil of the flat bodies lying in swales or near the water level is inclined to be mottled in color, clammy, and insufficiently aerated on account of poor drainage. The productiveness of the Sassafras fine sandy loam depends largely upon the drainage.

The type is confined largely to the necks and water fronts lying below the 25-foot contour line. The most important areas are those including the greater part of the neck south of Claiborne and the upland strip occurring along the 50-foot contour line or rim of the highland bordering the Choptank River valley from a point just south of Trappe to the neighborhood of Manadier. The topography of this higher area is moderately rolling. Its drainage is very good. The lower foreland is about equally divided between that having an undulating to slightly ridgy surface with good natural drainage and that having a nearly flat surface with poor underdrainage.

The type is fairly easy to cultivate, especially where the drainage is good, although in dry weather the soil is inclined to bake and harden in a way that makes fall breaking quite difficult, requiring considerable harrowing and rolling to bring it into a good tilth.

About 50 per cent of the Sassafras fine sandy loam is under cultivation, the remainder being forested mainly with pine, white oak. chestnut, sweet gum, black gum, and dogwood. It is devoted to general farming, including the production of wheat, corn, grass, and tomatoes. The yields of corn and wheat are about the same as on the Sassafras sandy loam, but the yields of timothy are thought to be better. Although the type does not have the texture of a typical truck soil, very fair yields of tomatoes of good quality are grown. Alfalfa seeded in the fall would do well on the better drained phase. Like the Sassafras loam, it requires regular applications of vegetable manure, barnyard manure, or legumes turned under in order to maintain favorable structural conditions.

The poorer drained areas are so situated that very effective drainage systems could be installed with inconsiderable outlay. Good results could be had with deep open ditches through the lower depressions; but much more satisfactory results would be obtained by installing lateral tile drains to discharge into main open ditches.

The following table gives the results of mechanical analyses of samples of the soil, subsoil, and lower subsoil of this type:

Mechanical analyses of Sassafras fine sandy loam.

Number	Description	Fine gravel	Coarse sand	Medium sand	Fine sand	Very fine sand	Silt	Clay
		Per cent	Per cent	Per cent	Per cent	Per cent	Per cent	Per cent
17927	Soil	0.1	2.0	14.3	29.2	10.1	35.6	8.7
17928	Subsoil	.2	1.6	12.8	28.0	8.0	25.9	13.8
17929	Lower subsoil .	.0	1.4	20.0	41.6	10.0	15.2	12.1

SASSAFRAS LOAM.

The surface soil of the Sassafras loam consists of a brown or yellowish-brown moderately heavy loam from 8 to 16 inches deep. The typical subsoil is a reddish-yellow heavy loam carrying about the same amount of silt but a little more clay than the soil. It becomes reddish brown in the lower portion, passing at 26 to 32 inches into a reddish-brown coarse sandy loam to sticky coarse sand with fine quartz gravel. Often the subsoil is pale yellow or yellow in the upper part of the profile, but always tends toward a reddish color with increase in depth. As a general thing the better the drainage the more nearly reddish brown becomes the subsoil. The type is locally called "red clay" or "medium stiff loam."

Not infrequently small quartz gravel is distributed throughout the soil mass, but never in sufficient quantity to influence materially the moisture-holding capacity. Minute flakes of mica can be seen distinctly in the subsoil in some places, though not so generally as to be a distinguishing feature.

Although the texture of the Sassafras loam is quite uniform, it requires close observation to determine the exact boundary between its lightest phase and the heavier Sassafras sandy loam, and, on the other hand, the gradation into the lighter phase of the Sassafras silt loam is also very insensible.

Both soil and subsoil have an ideally well-balanced texture; that is, the constituent particles of sand, silt, and clay are so proportioned and arranged as to constitute a soil capable of maintaining moisture in the most favorable quantity for the healthy growth of plants. However, where the organic content has been allowed to run down through continuous cropping without replenishment of vegetable matter, and where the land has stood long without cultivation—especially if cattle have been permitted to graze it closely and pack it during all kinds of weather—the soil assumes a hardened, compact structure unfavorable to cultivation, especially in dry falls. Such fields break up in clods and repeated rolling and harrowing are required to restore a good tilth. If the supply of humus is maintained and due attention is paid to moisture conditions in their relation to plowing and grazing, the soil can be kept in the best condition of tilth.

The Sassafras loam is confined largely to the uplands of Talbot County. A small percentage occurs in the lower foreland country, and it is even found along the steeper slopes. Its topography is quite like that of the Sassafras sandy loam, though generally somewhat less rolling. Where the relief has favored erosion the surface soil has been removed from small spots, revealing the yellowish-red subsoil in such a way as to give fields a spotted appearance. These spots, owing to the higher content of clay, are a little more difficult to cultivate than the true surface soil. They are sometimes called "clay hills."

While the texture affords good natural drainage, an occasional flat or depressed area will need main outlet ditches and tiling.

The Sassafras loam owes its origin to the weathering of marine deposits under good drainage conditions. It has since its uplift undergone about the same degree of weathering as the Sassafras sandy loam.

This is the best general farming soil of the county, being ideally adapted to wheat, corn, grass, clover, and forage crops and quite well suited to certain truck crops like tomatoes, beans, and cabbage. Wheat yields from 18 to 35 bushels, corn 40 to 75 bushels, and grass 1 ton to 2½ tons per acre. Where the soil is kept up to a good state of productiveness, as under a five-year rotation of corn, wheat, grass, wheat, grass, applying barnyard manure and 40 bushels of lime to the broken grass sod preceding corn and about 300 pounds of good commercial fertilizer to wheat, average yields of 60 bushels of corn, 20 bushels of wheat after corn and 28 bushels after grass, and 1½ tons of hay per acre are readily secured. If occasional crops of cowpeas or clover, to be turned under in conjunction with 40 bushels of lime per acre, should be introduced into the above general scheme of rotation, the most worn fields of the type could be brought quickly up to a point of equal productiveness or even better. Alfalfa does well when seeded properly; that is, in the fall, on a thoroughly prepared seed bed.

A large acreage of tomatoes is grown on this type, the average yield being about 4 tons per acre. Strawberries, cantaloupes, asparagus, beans, and buckwheat do well. Peaches, pears, and raspberries grow rapidly and bear well. In view of the good yields of hay and forage crops stock raising should be extended. Dairying and sheep raising could also be made quite profitable. There is no excuse for other than good average yields on the Sassafras loam, as it is easily improved and kept in good condition. Its valuation varies greatly with location.

The following table gives the average results of mechanical analyses of samples of the Sassafras loam:

Mechanical analyses of Sassafras loam.

Number	Description	Fine gravel	Coarse sand	Medium sand	Fine sand	Very fine sand	Silt	Clay
		Per cent	*Per cent*	*Per cent*	*Per cent*	*Per cent*	*Per cent*	*Per cent*
17003, 17933......	Soil	2.3	13.7	9.9	12.7	3.2	50.1	8.4
17004, 17934......	Subsoil	1.3	10.0	9.5	13.9	4.3	46.1	14.3
17005, 17935......	Lower subsoil ..	5.2	25.2	17.4	21.8	3.1	14.1	12.5

ELKTON SANDY LOAM.

The surface soil of the Elkton sandy loam consists of 6 to 10 inches of dark-gray, clammy, rather silty sandy loam which becomes light gray in the lower portion. The subsoil is a clammy, medium heavy sandy loam to loam carrying considerable silt. The upper portion is a light-gray to drab, frequently slightly mottled with reddish-yellow streaks, while the lower portion is generally intensely mottled with grayish, reddish-yellow, and reddish-brown colors. Strata of clayey material are quite common in the lower portion, and at a depth of about 30 inches occurs a substratum of compact, light-gray or mottled, sticky, medium to coarse sand which is always saturated with water.

The type occurs as small bodies in depressions and as low, flat land around the heads of small streams. The drainage is very poor, the water table generally standing very near the surface. The surface configuration of those areas found near the heads of streams is usually interrupted by small saucerlike depressions holding dark-colored material high in organic matter and generally in a semi-marshy condition. There are small bodies throughout the uplands of the area, many of which are so small as to necessitate their being included with other types.

The Elkton sandy loam is derived from the same material that gives rise to the Sassafras sandy loam. The original material, subjected to intermittent wet and dry stages, has undergone unfavorable structural and probably chemical changes and has accumulated a small amount of organic matter in an apparently stagnant, unhealthy condition.

Most of the Elkton sandy loam is forested with sweet gum, white oak, maple, black gum, dogwood, scattered pine, and a thick undergrowth ·of shrubbery. Under present conditions of drainage it is hard to manage and comparatively unproductive. Buckwheat, strawberries, and dewberries do fairly well where the drainage is best. By deepening, strengthening, and extending the natural drainageways and by putting in close lateral tiles or even open ditches, most of the type could be brought into pretty good condition without a prohibitive outlay of money. Many of the small depressions can be drained simply by a deep outlet ditch. Wheat, strawberries, dewberries, and grass would do quite well. Coarse barnyard manure plowed under improves the aeration considerably, under fair conditions of drainage. The type is not as desirable a soil as the Portsmouth sandy loam.

The following table gives the average results of mechanical analyses of samples of the soil and subsoil of this type:

Mechanical analyses of Elkton sandy loam.

Number	Description	Fine gravel	Coarse sand	Medium sand	Fine sand	Very fine sand	Silt	Clay
		Per cent	Per cent	Per cent	Per cent	Per cent	Per cent	Per cent
17892, 17894......	Soil	1.7	14 9	14.2	26.1	5 0	31.8	6.3
17893, 17895......	Subsoil	1.4	12.8	12.8	29.2	5.3	25.2	13.1

A determination of the organic matter gave the following percentage: No. 17892, 1.56 per cent.

ELKTON LOAM.

In color and profile the Elkton loam is very similar to the Elkton sandy loam. The surface soil from 6 to 10 inches consists of a dark-gray, silty, medium loam which becomes light gray in the lower portion. The subsoil to a depth of about 30 inches is a light-gray to drab, clammy, silty, heavy loam with a slight mottling of reddish yellow in the upper portion and intense mottling in the lower portion. Thin strata of clayey material occur in the lower profile. Below 30 inches there is usually found a substratum of light-gray, compact, medium to coarse sand always saturated with water.

The Elkton loam, like the Elkton sandy loam, occurs as poorly drained depressions and flat land around the heads of small streams and in small depressions holding darker colored, soggy soils.

There is a phase of the type occurring in small, nearly flat bodies on the neck of land to the south of Claiborne, and in such places the soil to an average depth of 10 inches is a light-gray to ashy-gray, quite silty, very fine sandy loam.

About 75 per cent of the type is timbered mainly with white oak, maple, and gum. In its crop adaptation it is quite like the Elkton sandy loam, producing, however, better grass and wheat, but not as good corn, strawberries, and dewberries. It is probable that red-top, clover, herd's-grass and millet would do well on this soil. The type under present conditions of drainage is too soggy and cold to constitute even a fair agricultural soil. By draining the soil proper could be deepened and the subsoil opened, increasing the circulation of air and extending the zone for root development. Turning under partially rotted, coarse manure in conjunction with about 50 bushels of lime to the acre would go far toward rectifying the condition of the soil, though little permanent benefit can be expected until the soil is thoroughly drained. For this purpose deep open ditches and tiling are recommended.

The following table gives the results of mechanical analyses of a sample of the soil and of the subsoil of the Elkton loam:

Mechanical analyses of Elkton loam.

Number	Description	Fine gravel	Coarse sand	Medium sand	Fine sand	Very fine sand	Silt	Clay
		Per cent	Per cent	Per cent	Per cent	Per cent	Per cent	Per cent
17896	Soil	2.0	12.7	11.0	16.5	3.0	48.9	6.4
17897	Subsoil	.9	10.4	10.0	15.1	3.1	49.7	10.0

PORTSMOUTH SANDY LOAM.

The Portsmouth sandy loam is a very dark-gray to black medium sandy loam of high organic-matter content, varying in depth from 8 to 12 inches. The subsoil to about 28 inches is a gray sandy loam with a slightly higher silt and clay content and much lower organic matter content than the soil. The subsoil frequently is mottled with reddish yellow and in the lower portion may carry strata and pockets of sandy clay. It usually passes into a light-gray or nearly white, compact, coarse sand to sticky sand, which also may contain pockets and thin strata of sandy clay. This sandy substratum is always saturated, except where thoroughly drained. It is sometimes washed out into open ditches, filling the bottoms and causing the banks to cave. In some places the soil may consist largely of organic matter mixed with just enough earthy material to give it the characteristics of Muck. Again, small areas may consist, from a few inches to a foot or more, of bog iron ore, occurring on top or at any position in the profile.

Uncleared areas support a growth of sweet gum, black gum, beech, maple, and scattering pine, with a dense undergrowth of whortleberry, gallberry, and other bushes.

The Portsmouth sandy loam is confined to a very small portion east of Wye Mills.

The type is derived from the same material that gives rise to the Sassafras and Elkton sandy loams. The toporaphy has induced wet, swampy conditions which have favored accumulation and preservation of a large quantity of spongy vegetable matter in the soil, at the same time retarding subsoil weathering.

When thoroughly drained the Portsmouth sandy loam proves a very productive soil. Excellent yields of strawberries, corn, and dewberries are made. Owing to a tendency of the soil upon freezing to heave, the type is not especially suited to fall-sown crops. Wheat and grass are particularly likely to suffer in this way. It is claimed, however, that wheat grown on this soil is of the good hard quality, preferred by millers. Buckwheat and oats would do well, as would onions and celery also. Tomatoes do only fairly well. Rather heavy applications of lime—from 40 to 60 bushels per acre—at frequent intervals are required to bring the soil up to its maximum producing capacity. Deep plowing is said to be quite beneficial.

The following table gives the results of mechanical analyses of samples of the soil, subsoil, and lower subsoil of the Portsmouth sandy loam:

Mechanical analyses of Portsmouth sandy loam.

Number	Description	Fine gravel	Coarse sand	Medium sand	Fine sand	Very fine sand	Silt	Clay
		Per cent	*Per cent*	*Per cent*	*Per cent*	*Per cent*	*Per cent*	*Per cent*
17910	Soil	1.2	24 2	26.0	26.5	1.9	12.5	7.3
17911	Subsoil	1.1	16.4	20.8	35 5	5.1	12.7	7.7
17912	Lower subsoil	2.0	17.0	20.7	45.6	5.0	7.0	2.7

MEADOW.

Wet alluvial bottom land having no uniformly definite texture and those areas near the heads of streams subjected throughout the year to standing water and swampy conditions have been classed as Meadow. A considerable proportion of the alluvial phase is susceptible to easy reclamation, and some of it is under cultivation. The upland bodies can be reclaimed by clearing, straightening, and extending the natural drainage ways. A few small areas are found along the water front just above tidal overflow. Generally every stream and drainage way has a narrow strip of Meadow from its mouth to its source. When thoroughly drained Meadow produces excellent corn and, where the organic-matter content is not too high, good grass and wheat.

TIDAL MARSH.

The areas of marshy land lying near water level and subject to tidal overflow are classed as Tidal marsh.

This land is a black or brown slimy loam, clay loam, or ooze, which is generally underlain at about 2 feet by a sandy or clayey material. The mass is pretty thoroughly interspersed with the roots of coarse grasses and cattails and decomposing vegetable matter.

The whole remains saturated the year around. Considerable hydrogen sulphide is developed in the lower portion by decomposing vegetable matter. The odor can be very distinctly detected in the freshly exposed material, particularly in those areas touching salt water.

Tidal marshes occur as narrow fringes along the larger bodies of water. The most extensive areas are found along the upper Choptank River.

The material is sedimentary in origin, having been deposited by rivers or built up by tides and waves. It can be reclaimed only by diking to keep out the tides. It supports a rank growth of coarse grasses and cattails. The grass is sometimes cut for hay.

THE CLIMATE OF TALBOT COUNTY

BY

ROSCOE NUNN

INTRODUCTORY.

Geographical situation and topography being prime factors of climate, a brief description of the region to be discussed is first in order. Talbot County is the smallest of the Eastern Shore counties of Maryland, and it has a larger percentage of its area in mixed land-and-water surface than any other county. Its total area is 267 square miles, and of this about one-fourth is in the form of peninsulas and islands in the Chesapeake Bay. With its extreme western limit, Poplar Island, Talbot reaches farther west than any other county of the Eastern Shore. Its chief city, Easton, latitude north 38° 46′ and longitude west 76° 5′, is situated almost at the center of the county, yet is within two miles of tide water.

Easton is directly southeast of Baltimore, at an air-line distance of 50 miles. It is the only point in the county at which a long series of weather observations and records have been made; but this station is ideally located to represent the general climatic characteristics of the county. Its general elevation above sea level is about 30 feet, and this elevation is not far from the average elevation of the county, as there are but few points higher than 60 feet above sea level.

The general trend of the county boundaries is northeast-southwest and the drainage is from the northern parts towards the south and southwest. The county is bordered by water on all sides except its short northern side; the Wye River being on the northwest, the Chesapeake Bay on the west, and the Choptank River on the south and southeast.

The physical setting of Talbot County—its marine aspect and its rather uniform topography—has a distinct influence upon its climate, the chief characteristics of which are mildness and freedom from great extremes. These characteristics may be easily recognized when comparison is made of Easton climatic records with those of places of the same latitude in the interior, where increased elevation above sea level and the absence of the tempering effects of water areas give such places greater ranges and extremes of temperature than are experienced on the Eastern Shore of Maryland.

CLIMATIC RECORDS AVAILABLE.

There is an excellent series of records for Easton, beginning November 1, 1891, and continuing to the present time without a break, except that during the first year the records are not complete. These records were kept, first, by Mr. S. P. Minnick, November, 1891 to November, 1892; then by Mr. George W. Minnick, December, 1892 to May, 1894. On June 1, 1894, Mr. Henry Shreve took charge of the weather station, and he kept an unbroken record until the time of his leaving the State, October 25, 1922, a period of 28 years and 4 months. Mr. Shreve was succeeded by Mr. Clement E. Bray, the present efficient observer. Although not available to this office, a series of records, it is understood, was kept by Capt. Jacob G. Morris, who began recording the minimum temperature about 1867 and the maximum temperature about 1888. This private record was kept for many years.

The records by Messrs. Minnick, Shreve, and Bray were made in co-operation with the United States Weather Bureau. Standard instruments were used and approved methods of exposure and reading of the instruments were employed. The instruments and other equipment were furnished by the Weather Bureau and the records of the observers have been carefully scrutinized and supervised by Weather Bureau officials. There was no material change in the location of the instruments during the 34 years covered.

DISCUSSION OF THE DATA.

Statistics are essential in a correct description of climate. It is only by actual records for a long period of years that the characteristics of the climate of any region can be set forth. And it is only by comparison of the climatic statistics of a particular region with those of other regions that the relative advantages of one climate over another can be seen.

In the accompanying diagrams and tables a fairly complete exhibit of the climate of Talbot County is found. For comparative data of other portions of Maryland, or for other sections of the United States, the reader is referred to the publications of the Weather Bureau, which may be procured from the central office at Washington or from the Weather Bureau station most convenient to the applicant.

TEMPERATURE.

Table I shows the average, or mean, temperature conditions for each month and closes with a line showing the general averages, or normals, obtained from the whole record. Here may be seen all the fluctuations of temperature during a 34-year period; the coldest month and the warmest; the ranges of mean temperature in the history of any month, season, or year. For example, the coldest January was that of 1893, with an average temperature of 23.9°, while the warmest January was that of 1913, with an average of 44.8°. Thus we see that one January may be 21 degrees warmer or colder than another; and a particular January may be 10 degrees warmer or colder than the normal. In the summer months it is seen that the range is much less; for in July the highest average was 79.8°, in 1901, and the lowest, 72.0°, in 1895, showing a difference of only 7.8° between the extremes.

The progress of the seasons can be seen from Table I; how the coldest time of the year is in January and February, how March averages about 10 degrees warmer than February, April 9 degrees warmer than March, and May 10 degrees warmer than April; then comes a more gradual approach to the peak of summer in July. Table I shows that the seasons in Talbot County are well marked, a most enjoyable feature of the climate. Figure 1 also illustrates the progress of the seasons.

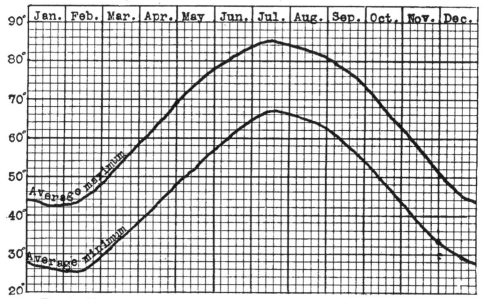

FIG. 1.—33-YEAR AVERAGE MAXIMUM AND MINIMUM TEMPERATURES THROUGH THE YEAR AT EASTON.

General averages, such as we have just been discussing, are not at all sufficient to describe climate. We therefore give in tables II III, and IV much more details of temperature. Table II shows the averages of the highest and lowest daily temperatures. We see here the temperature zone in which Talbot County finds itself from day to day and month to month; for example, during January days the temperature ranges on an average, from a minimum of about 26.6° to a maximum of about 42.6°. In July the zone lies between 67.1° and 85.1°. This average zone is graphically shown in figure 1.

Extremes of temperature—that is, the daily highest and lowest temperatures—are an important feature of climate. Much information thereon is given in tables III and IV. In Table III will be seen the highest temperature for each month of record and in Table IV the lowest for each month. We find here only moderate extremes as compared with the extremes experienced in the interior or mountain regions of Maryland and other States. The highest recorded at Easton in 34 years was 101 degrees and the lowest 15 degrees below zero.

THE GROWING SEASON.

The length of the growing season for each year, or the dates of the last freezing temperature in spring and the first in autumn, with the number of days between, is shown in Table V. Crops are safe from frost, on an average, from about April 10 to October 29, but killing frost has occurred, once, as late as April 28 and as early as October 2; very few times in the last 34 years, however, has killing frost occurred later than April 15 or earlier than October 20.

PRECIPITATION.

Precipitation records are rather fully set forth in tables VI, VII, VIII, and IX. Close examination of these tables will reveal the principal facts in regard to this important feature of the climate of Talbot County. Rainfall is quite a variable element of climate over most of the United States. In Talbot County we find a rather uniform distribution of rainfall through the year; that is, the normal amounts show no great range from month to month or from season to season. Particular months, seasons, and years show marked variations from normal.

It is noteworthy that the rainfall is ample during the crop growing season, the July normal being the highest of the year, with August coming second and June third. The driest season is autumn. For the year as a whole, rainfall and snowfall are moderate in Talbot County, when compared with precipitation in the best agricultural regions of the United States.

Figure 2 and the rainfall tables show striking variations in monthly and annual precipitation. For example, July has had as little as 1.47 inches, in 1914, and as much as 9.03 inches, in 1919. The driest year was 1896, with 30.93 inches, and the wettest, 1919, with 55.06 inches. In the entire record of 34 years there has been no month without appreciable rainfall. The smallest monthly amount was 0.21 inch, in May, 1911.

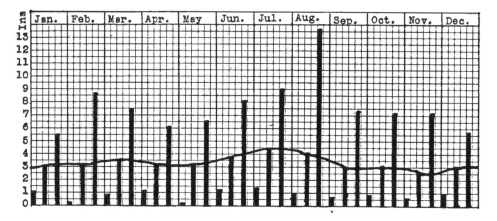

FIG. 2.—MONTHLY MEAN AND EXTREMES OF PRECIPITATION AT EASTON.

Table VIII gives an interesting history of snowfall at Easton during the 34 years of record. Easton, considering its latitude, has a light average annual snowfall.

The frequency of occurrence of precipitation is well shown in Table IX, which gives the number of days for each month on which measureable precipitation occurred. This record shows how uniformly rain falls during the different seasons, on an average. It also shows how much rainier some particular months may be than others. The average number of days with precipitation per month in winter is 8; in spring and summer, 9; in autumn, 6.

OTHER FEATURES OF THE CLIMATE.

Temperature, frequency of certain extremes.—The average number of days in the year with maximum temperature as high as 90°

or above is 15; with minimum temperature as low as 32° or below, 92; with minimum as low as 14° or below, 9. Temperatures of zero or lower, seldom occur; in only 7 out of the 34 years of record has zero or lower been registered. On an average, zero temperature occurs about once in five years.

Sunshine and cloudiness.—Talbot County receives about the same percentage of sunshine as the Baltimore district, or about 60 per cent of the possible amount. This means an average of about 120 days clear, 127 partly cloudy, and 118 cloudy.

Humidity.—While Talbot County station was not equipped for making observations of relative humidity, and no actual records are available, it is evident from the records of the nearest regular Weather Bureau stations, especially Baltimore, that the average relative humidity for the region is about 70 per cent. This is slightly lower humidity than prevails on the immediate Atlantic coast and slightly higher than that of the Piedmont and mountain regions of Maryland.

Excessive rainfall.—Records of daily rainfall show the number of occurrences of excessive amounts (that is, 2.50 inches or more within 24 hours). It appears that falls of 2.50 inches or more in a day may be expected less than once a year, on an average. At Easton such excessive falls occurred 22 times in 34 years. In 21 of the 34 years there was no excessive fall. The greatest number in any year was 3 in 1919. Excessive rains occur principally in the summer months, with thunderstorms, but some occur in connection with the Atlantic coast storms of September and October. The greatest amount on record for 24 hours is 6.00 inches. Amounts exceeding 4.00 inches in 24 hours seldom fall.

Thunderstorms.—Thunderstorms occur principally in the months of June, July, and August. The total number for the year, on an average, is 36. This is less than half the annual number in the lower Southern states.

Tornadoes.—There is no authentic record of a tornado on the Eastern Shore. Talbot County shares in this immunity.

Prevailing winds.—During the winter and spring the winds come mostly from the northwest or southwest. In the summer south and southwest winds predominate. During the autumn the winds are most frequently from the southwest, but north, northwest, and south winds are next in order of frequency. The least frequent winds are those from easterly quarters, but the easterly winds are particularly notable because they are so often attended by cloudiness and precipitation, and, in summer, by lowered temperature.

TABLE I.

MONTHLY AND ANNUAL MEAN TEMPERATURES AT EASTON.

Year	Jan.	Feb.	March	April	May	June	July	Aug.	Sept.	Oct.	Nov.	Dec.	Annual
1891		38.	39.	54.	64.8	6.	76.6	2			44.3	43.0	
1892	34.3	4.	41.	55.	63.6	3.	77.2	.6	.5	56.4	42.8	30.1	54.5
1893	23.9	7.	50.	5	67.5	2.	77.3	.6	.6	58.4	44.0	38.2	54.3
1894	39.0	8.	42.	5	62.2	2.	72.0	.0	.6	60.0	45.4	40.9	57.4
1895	35.2	6.	39.	5	67.1	9.	76.9	.5	.6	51.8	47.2	39.6	54.5
1896	32.5	5.	44.	5	61.9	8.	76.0	.4	.4	54.0	51.0	34.5	55.0
1897	30.9	4.	47.	4	62.4	2.	78.4	.6	.8	57.3	45.9	37.6	54.3
1898	36.0	7.	42.	5	62.8	4.	76.8	.6	.5	59.0	43.2	35.2	55.5
1899	33.0	3.	39.	5	63.2	2.	79.5	.6	.4	58.2	44.8	36.6	54.2
1900	36.2	9.	44.	5	61.4	8.	79.8	.5	.8	61.8	49.3	36.0	56.4
1901	33.4	29.	46.	5	64.4	2.	76.6	.6	.1	56.4	41.9	35.6	53.9
1902	32.0	38.	51.	5	64.3	7.	76.2	.0	.8	59.1	51.6	36.2	55.3
1903	33.6	29.	42.	5	64.4	0.	74.6	.8	.4	53.4	43.2	30.6	54.9
1904	28.5	26.	44.	5	64.9	1.	75.6	.9	.6	57.4	45.3	38.6	52.3
1905	30.4	35.	38.	54.	63.8	2.	74.2	.0	.6	57.0	46.6	37.2	54.0
1906	40.2	29.	46.	47.	58.6	5.	75.2	.7	.7	52.7	46.4	40.2	55.6
1907	37.8	32.	47.	55.	65.0	0.	77.8	.6	.2	58.7	46.8	37.4	53.4
1908	35.8	43.	42.	53.	62.8	1.	72.9	.4	.6	53.4	49.7	32.5	55.3
1909	37.7	35.	49.	56.	61.2	8.	75.7	.0		60.0	42.2	30.8	54.7
1910	33.8	36.	41.	48.	67.2	2.	78.0	.6		57.0	45.2	41.7	54.5
1911	38.6	32.	41.	55.	64.6	0.	75.2	.8	.7	59.5	48.2	40.5	53.9
1912	26.3	36.	49.	55.	63.2	0.	76.9	.0	.6	60.2	48.4	40.3	55.8
1913	44.8	31.	38.	52.	65.4	2.	74.5	.7	.0	61.2	47.0	34.2	57.2
1914	38.1	39.	38.	57.	62.4	9.	75.6	.8	.8	59.4	47.5	35.2	54.7
1915	37.2	34.	37.	52.	66.0	9.	77.2	.1	.1	58.4	47.5	36.4	55.5
1916	40.2	33.	43.	53.	59.0	2.	76.4	.8	.7	53.8	43.0	29.6	55.1
1917	36.0	35.	47.	53.	68.6	0.	74.6	.6	.2	61.0	47.2	42.0	53.3
1918	24.0	37.	47.	53.	64.9	2.	76.1	.2	.6	64.0	48.4	32.6	55.5
1919	38.4	32.	44.	53.	58.8	70.	74.0	.8	.6	61.7	46.4	40.4	56.4
1920	29.8	39.	55.	58.	61.6	73.	79.0	.2	.8	57.6	49.3	37.6	54.7
1921	38.3	38.	45.	55.	64.9	73.	75.2	.6	.7	60.0	49.2	39.4	58.2
1922	32.0	33.	46.	53.	63.7	75.	74.9	.8	.6	57.2	46.2	45.6	56.2
1923	37.9	35.	43.	51.	61.0	71.	74.7	.4	.5	57.0	46.8	38.2	56.6
1924	37.0	44.	47.	56.	61.0	77.	76.8	.9	.4	53.1	45.2	37.2	54.7
1925	33.9												56.4
Av.	34.6	34.6	44.3	53.6	63.5	71.4	76.1	74.2	68.6	57.8	46.3	37.0	55.2

TABLE II.
MEAN MAXIMUM AND MEAN MINIMUM TEMPERATURES.
Mean Maximum

Year	Jan.	Feb.	March	April	May	June	July	Aug.	Sept.	Oct.	Nov.	Dec.	Annual
	42.6	42.9	54.2	64.0	73.9	80.9	85.1	83.2	78.4	67.8	55.6	46.2	64.6

Mean Minimum

| | 26.6 | 26.2 | 34.8 | 43.2 | 53.0 | 61.6 | 67.1 | 65.1 | 59.0 | 47.8 | 37.3 | 29.2 | 45.9 |

TABLE III.
HIGHEST TEMPERATURES AT EASTON.

Year	Jan.	Feb.	March	April	May	June	July	Aug.	Sept.	Oct.	Nov.	Dec.	Annual
1892	62	60	....	....	....	....	....	....	....	....	....	67	....
1893	52	63	66	80	87	95	95	92	88	82	64	62	95
1894	57	63	82	80	87	96	95	89	92	80	72	64	96
1895	61	64	74	84	93	95	93	95	93	74	77	66	95
1896	59	63	69	93	91	88	92	98	94	77	74	60	98
1897	65	56	75	85	83	93	90	89	93	86	72	65	93
1898	63	63	71	81	89	96	101	95	96	87	67	63	101
1899	59	60	70	82	89	95	93	95	93	83	67	65	95
1900	60	65	74	76	91	91	99	101	94	88	78	60	101
1901	60	53	77	79	80	94	99	91	91	80	65	68	99
1902	52	68	76	85	87	91	97	89	89	77	73	70	97
1903	55	69	73	88	93	87	96	97	90	84	74	52	97
1904	58	58	72	83	87	94	92	89	91	84	63	61	94
1905	61	49	78	82	85	90	93	88	86	84	69	59	93
1906	71	62	60	83	90	91	89	89	88	77	72	65	91
1907	70	53	89	77	79	89	90	88	89	77	64	66	90
1908	60	62	75	83	87	93	96	91	81	84	72	69	96
1909	62	71	72	82	86	91	90	89	84	77	75	60	91
1910	57	71	80	83	84	89	91	89	90	84	76	59	91
1911	60	66	71	74	92	94	96	93	86	79	69	66	96
1912	55	59	69	76	83	88	90	91	91	82	74	67	91
1913	67	69	78	81	88	92	94	93	91	78	75	61	94
1914	70	62	72	80	91	93	94	95	92	84	77	61	95
1915	61	67	58	89	83	86	93	96	91	77	73	62	96
1916	69	63	67	82	89	85	91	95	90	84	73	67	95
1917	58	64	76	85	86	92	94	93	85	79	67	53	94
1918	58	67	77	77	90	93	93	100	84	82	71	66	100
1919	61	67	73	76	88	89	95	88	88	88	73	65	95
1920	60	53	74	80	82	91	93	88	86	83	72	67	93
1921	62	69	84	83	85	95	95	94	94	78	77	65	95
1922	57	71	74	85	84	90	92	86	87	85	71	64	92
1923	62	58	81	82	89	97	96	96	89	83	64	67	97
1924	61	64	77	80	85	95	95	100	93	78	72	72	100
1925	60	70	76	85	96	100	96	92	94	80	70	61	100
Highest	71	71	89	93	96	100	101	101	96	88	78	72	101

TABLE IV.
LOWEST TEMPERATURES AT EASTON.

Year	Jan.	Feb.	March	April	May	June	July	Aug.	Sept.	Oct.	Nov.	Dec.	Annual
1892	10	13	....	....	....	....	....	....	....	....	....	17	10
1893	—1	11	15	37	43	53	58	57	44	31	23	17	—1
1894	22	14	23	30	47	45	52	51	45	37	21	12	12
1895	11	2	23	30	39	51	54	52	43	28	25	16	2
1896	8	7	18	27	38	50	55	50	38	29	28	12	7
1897	8	15	21	30	43	40	61	53	39	34	24	15	8

TABLE IV.—Continued.
LOWEST TEMPERATURES AT EASTON.

Year	Jan.	Feb.	March	April	May	June	July	Aug.	Sept.	Oct.	Nov.	Dec.	Annual
1898	17	10	24	26	38	51	54	57	43	30	26	13	10
1899	6	—15	24	27	40	53	52	57	41	31	25	8	—15
1900	12	7	11	27	35	49	56	55	45	33	26	12	7
1901	11	14	14	35	43	50	66	55	45	32	21	9	9
1902	17	8	21	31	42	49	59	51	43	30	27	16	8
1903	12	6	23	28	34	45	54	52	39	32	15	11	6
1904	3	2	18	27	42	46	54	51	35	28	23	0	0
1905	—7	—3	17	29	39	48	60	52	42	31	17	17	—7
1906	12	8	20	29	40	51	54	63	49	31	25	12	8
1907	8	5	20	23	37	48	57	50	45	30	27	20	5
1908	10	6	24	28	39	47	59	51	41	31	25	8	6
1909	12	17	22	25	37	51	52	51	41	33	27	9	9
1910	12	8	21	33	37	47	50	51	41	31	21	7	7
1911	18	20	8	27	35	50	55	54	44	36	25	23	8
1912	—8	7	17	29	41	45	54	52	42	39	23	18	—8
1913	24	13	17	31	33	45	55	56	42	35	29	21	13
1914	6	2	11	27	40	48	55	50	38	35	21	7	2
1915	18	19	23	30	42	49	57	53	41	35	24	20	18
1916	5	4	14	33	44	48	54	56	41	34	24	8	4
1917	15	2	23	27	38	54	62	51	40	27	1	—3	—3
1918	2	—4	22	31	42	51	50	52	42	33	30	22	—4
1919	14	22	28	25	45	49	52	57	45	40	26	3	3
1920	9	7	15	30	35	55	53	56	44	41	21	20	7
1921	10	20	27	27	40	45	60	50	53	34	27	11	10
1922	6	7	23	30	37	53	56	54	43	30	28	17	6
1923	18	14	18	16	38	51	53	49	40	35	24	23	14
1924	9	17	26	28	40	50	57	52	42	30	19	13	9
1925	4	19	12	28	36	55	52	49	40	29	23	13	4
Lowest	—8	—15	8	16	33	40	50	49	35	27	15	—3	—15

TABLE V.
KILLING FROSTS AT EASTON.

Year	Last in Spring	First in Autumn	Length of Growing Season, Days
1893:	March 30	October 31	214
1894:	April 3	November 12	222
1895:	April 12	October 10	180
1896:	April 9	October 22	195
1897:	April 21	November 14	207
1898:	April 28	October 28	183
1899:	April 11	October 2	174
1900:	April 11	November 15	218
1901:	March 30	October 26	209
1902:	April 16	October 30	197
1903:	April 6	October 28	205
1904:	April 23	October 28	188
1905:	April 19	October 22	186
1906:	April 3	October 13	193
1907:	April 20	October 23	185
1908:	April 17	November 5	202
1909:	April 12	November 19	221
1910:	March 22	October 31	223
1911:	April 11	November 3	206
1912:	April 9	November 4	209
1913:	April 9	November 1	208
1914:	April 11	November 7	210
1915:	April 5	November 4	213
1916:	March 25	November 4	224
1917:	April 15	October 13	181
1918:	April 6	November 3	211
1919:	April 2	November 10	222
1920:	April 11	November 13	216
1921:	April 12	November 6	208
1922:	April 3	October 21	201
1923:	April 10	November 2	206
1924:	April 3	October 23	203
1925:	April 21	October 20	182
Average	April 10	October 29	204

TABLE VI.
MONTHLY AND ANNUAL PRECIPITATION IN INCHES AT EASTON.

Year	Jan.	Feb.	March	April	May	June	July	Aug.	Sept.	Oct.	Nov.	Dec.	Annual
1891											1.19	2.33	
1892	4.69	.00	.9	.00	.5	.00	2.00	1.00	.8	.9	.00	2.32	36.79
1893	2.4	.	.4	.	.6	.4	4.	4.	.1	.4	.	2.90	39.89
1894	2.4	.	.8	.	.6	.3	5.	4.	.9	.0	.	2.60	41.78
1895	4.0	.	.4	.	.7	.1	4.	3.	.4	.0	.	1.72	37.07
1896	1.8	.	.3	.	.5	.7	2.	1.	.6	.4	.	0.98	30.93
1897	2.04	.	.1	.	.2	.3	8.	3.	.7	.2	.	4.12	43.71
1898	2.83	.	.8	.	.4	.4	2.	4.	.3	.96	.	3.96	36.99
1899	3.03	.	.7	.	.09	.1	6.	4.	.3	.70	.	1.42	43.62
1900	2.1	.	.1	.	.91	.2	2.	4.	.8	.5	.	2.49	42.48
1901	3.8	.	.1	.	.63	.9	5.	5.	.6	.8	.	5.76	41.00
1902	4.1	.	.6	.	.62	.3	3.	2.	.9	.1	.	4.58	50.64
1903	3.1	.	.4	.	.58	.1	4.	4.	.2	.5	.	2.55	47.17
1904	1.7	.	.4	.	.23	.8	4.	1.00	.9	.2	.	3.51	33.78
1905	3.6	.	.5	.	.51	.7	6.	3.97	.6	.7	.	3.49	39.40
1906	2.0	.	.7	.	.15	.0	5.	4.71	.2	.1	.	2.95	43.44
1907	1.3	.24	.3	.04	.53	.2	3.	3.62	.4	.3	.	3.45	52.08
1908	2.2	.05	.5	.83	.66	.6	4.	7.60	.0	.2	.	4.52	39.58
1909	2.3	.1	.3	.00	.09	.8	4.	2.53	.6	.2	.	3.01	39.65
1910	3.1	.8	.9	.	.99	.3	2.	3.82	.4	.4	.	2.32	39.82
1911	2.6	.6	.9	.	.21	.8	2.	13.72	.8	.3	.	3.27	44.32
1912	2.7	.4	.5	.	.30	.6	5.	1.86	.0	.8	.	3.39	40.76
1913	2.6	.8	.2	.	.69	.1	2.	1.25	.7	.9	.	2.23	34.70
1914	3.8	.6	.7	.	.16	.1	1.	3.38	.6	.3	.	4.12	31.67
1915	3.7	.6	.8	.	.24	.2	2.	7.00	.3	.8	.	3.05	37.53
1916	1.6	.7	.2	.	.19	.3	6.	1.	.2	.0	.	4.45	39.34
1917	2.3	.5	.1	.	.67	.8	5.	3.	.1	.8	.	1.41	38.03
1918	3.9	.9	.5	.	.76	.8	5.	1.	.7	.7	.	3.26	36.50
1919	3.3	.7	.7	.	.41	.8	9.	9.	.9	.3	.	3.51	55.06
1920	2.3	.4	.1	.	.41	.7	3.	7.	.4	.5	.	2.42	43.95
1921	2.1	.0	.5	.	.56	.2	4.	1.	.5	.9	.	1.85	31.13
1922	5.7	.0	.0	.	.58	.4	5.	3.	.0	.4	.	4.58	39.98
1923	4.2	.7	.9	.	.83	.5	7.	2.	.8	.6	.	3.48	44.01
1924	3.8	.04	.4	.	.29	.9	2.	5.	.2	.9	.	2.74	48.56
1925	4.9	.45	.1	.	.27	.7	6.	6.	.	.4	.	3.24	39.89
Av.	3.09	3.11	3.66	3.29	3.29	3.78	4.51	4.15	3.00	3.12	2.60	3.09	40.69

TABLE VII.
EXTREMES OF PRECIPITATION AT EASTON.
Greatest Monthly and Annual

5.57	8.61	7.54	6.13	6.53	8.13	9.03	13.72	7.43	7.34	7.35	5.76	55.06

Least Monthly and Annual

1.16	0.31	0.84	1.11	0.21	1.23	1.47	1.00	0.66	0.79	0.54	0.98	30.93

Greatest in 24 Hours

2.40	3.06	2.65	2.40	2.53	6.00	3.81	5.02	4.62	5.13	2.18	2.00	6.00

TABLE VIII.
MONTHLY AND ANNUAL UNMELTED SNOWFALL, IN INCHES, AT EASTON

Year	Jan.	Feb.	March	April	May	June	July	Aug.	Sept.	Oct.	Nov.	Dec.	Annual
1892	8.2	5.0										0.3	
1893	8.5		8.0	0						0	0	5.0	
1894	0		0	0						0	0	1.0	
1895	9.0	17.0	0	0						0	0	0.5	26.5

TABLE VIII.—Continued.

Year	Jan.	Feb.	March	April	May	June	July	Aug.	Sept.	Oct.	Nov.	Dec.	Annual
1896	0	0.5	1.5	T						0	0.5	2.5	5.0
1897	5.5	4.0	0	0						0	T	T	9.5
1898	1.0	T	0	1.0						0	0.7	1.2	3.9
1899	4.0	50.3	10.0	0						0	0	2.0	66.3
1900	0	10.0	....							0	0	7.2	....
1901	8.0	T	0	0						0	0.1	0.5	8.6
1902	8.0	5.0	T	0						0	0	2.4	15.4
1903	2.5	0.5	0	0						0	T	3.9	6.9
1904	4.2	2.8	....	0						0	....	7.0	....
1905	16.3	1.7	0	0						0	0	1.0	19.0
1906	1.2	3.7	2.6	0						0	T	T	7.5
1907	2.2	10.5	4.6	3.5						0	0	T	20.8
1908	5.1	1.0	0	0						0	0	7.0	13.1
1909	4.0	3.3	5.0	0						0	T	5.5	17.8
1910	1.4	3.0	5.6	0						0	T	9.6	19.6
1911	2.0	4.9	6.0	0						0	0	0	12.9
1912	10.4	0	6.1	0						0	2.7	4.8	24.0
1913	0	0.5	0	0						0	0	0	0.5
1914	0	6.4	14.1	0						0	0	1.0	21.5
1915	0.5	4.2	0	10.0						0	0	5.2	19.9
1916	4.5	4.0	6.5	0						0	T	7.5	22.5
1917	0.4	1.2	1.0	1.0						0	2.0	1.5	7.1
1918	16.8	1.0	0	0						0	0	T	17.8
1919	0	0.5	T	T						0	0	5.2	5.7
1920	1.4	7.8	3.3	T						0	0	T	12.5
1921	T	3.8	0	1.0						0	0	4.0	8.8
1922	27.5	7.2	T	0						0	T	1.6	36.3
1923	1.5	7.0	4.0	T						0	0	2.0	14.5
1924	T	6.0	4.2	1.5						0	T	0	11.7
1925	11.5	0	T	0						2.5	0	T	14.0
Av.	4.9	5.4	2.7	0.5						0.1	0.2	2.6	16.4

TABLE IX.

NUMBER OF DAYS WITH .01 INCH OR MORE OF PRECIPITATION AT EASTON.
(Rainfall and Melted Snow)

Year	Jan.	Feb.	March	April	May	June	July	Aug.	Sept.	Oct.	Nov.	Dec.	Annual
1892	11	10	...									6	....
1893	7	7	9	8	9	5	8	7	8	6	6	5	85
1894	6	7	5	10	13	5	9	6	7	8	7	6	89
1895	13	5	10	7	10	9	10	7	4	4	4	8	91
1896	3	7	11	6	11	12	9	4	8	4	5	4	84
1897	8	10	8	9	11	8	13	11	4	11	10	10	113
1898	10	5	10	10	13	7	6	9	4	13	10	8	105
1899	12	13	11	3	9	7	11	10	9	5	6	6	102
1900	5	7	11	7	7	8	7	6	10	8	7	5	88
1901	7	1	10	9	12	8	13	9	7	2	5	8	91
1902	7	5	10	8	5	9	9	5	5	7	7	9	86
1903	9	8	8	7	5	10	7	10	5	5	2	7	83
1904	8	5	9	6	4	12	10	4	3	2	4	9	76
1905	7	8	9	7	9	10	13	5	8	3	5	7	91
1906	11	7	12	7	5	16	12	17	4	12	7	5	115
1907	10	5	8	9	14	11	10	8	7	5	13	5	105
1908	10	9	8	8	12	6	9	10	7	5	5	10	99
1909	9	12	9	11	11	10	8	4	5	4	8	6	97
1910	8	5	4	11	13	12	11	12	5	7	7	8	103
1911	10	9	12	10	3	8	9	15	8	7	11	8	110
1912	10	7	11	11	8	9	10	7	9	3	3	8	96
1913	8	4	10	6	9	7	7	5	5	9	6	7	83
1914	6	8	13	9	5	11	5	6	2	5	3	11	84
1915	12	7	2	4	10	6	8	15	3	8	5	6	86
1916	8	8	8	11	10	9	6	4	5	2	9	9	89 .
1917	10	8	12	10	9	7	15	8	6	10	4	6	105
1918	9	6	9	•8	6	6	7	7	6	5	5	10	84
1919	7	8	10	6	13	5	12	12	4	8	9	10	104
1920	9	9	9	10	8	9	9	15	4	2	9	7	100
1921	5	10	10	11	11	4	7	6	7	3	14	7	95
1922	10	11	8	8	9	12	12	10	4	7	3	12	106
1923	12	9	13	8	5·	8	12	11	9	5	9	10	111
1924	7	7	10	12	16	12	4	8	11	2	5	10	104
1925	12	7	7	11	8	5	10	8	4	14	8	9	103
Av.	9	7	9	8	9	9	9	9	6	6	7	8	96

THE HYDROGRAPHY OF TALBOT COUNTY

BY

B. D. WOOD

Talbot County is located on the east shore of Chesapeake Bay and, with the exception of a small portion of the northern boundary, it is entirely surrounded by tide water as the fall of Choptank River and its tributary, Tuckahoe Creek, is so small that they are in the influence of tide to above the point where Tuckahoe Creek enters Queen Anne's County.

The Maximum elevation in the county exceeds 60 feet in only a few localities. Therefore the fall of the various streams is small and most of them are in the influence of tide water, nearly to their source, the extreme range of tide along the shores of the county being slightly over two feet.

No data are available in regard to the flow of the streams of the county, but judging from their size and the rainfall, which is 44 inches at Cambridge and 48 inches at Easton, the total flow of the various streams is necessarily small and as the slope is small the power possibilities in the county are limited, if any.

The following is a gazetteer of the streams of the county, not including the tidal estuaries. This gives a brief description of each stream as taken from the general county map published in 1905 by the Maryland Geological Survey in co-operation with the United States Geological Survey.

GAZETTEER OF STREAMS IN TALBOT COUNTY.[1]

Beaverdam Branch rises in Chapel township, Precinct No. 4, at altitude 60 feet above sea level; flows in general southeastward into

---

[1] Except where otherwise stated the only authority for the descriptions given is the county map. The gazetteer does not include the so-called creeks and rivers which are merely tidal estuaries.

Kings Creek (tributary to Choptank River, which discharges to Chesapeake Bay) ; length, about 5 miles; fall, about 50 feet.

Bolingbroke Creek rises in Trappe township, at altitude about 40 feet above sea level; flows in general southward into Choptank River, which discharges to Chesapeake Bay; length, about 4½ miles of which 2½ miles is a tidal estuary.

Choptank River rises in Kent County, Delaware, at altitude about 60 feet above sea level; flows southwestward into Chesapeake Bay; tidal for much of its course; principal tributary, Tuckahoe Creek, which forms the boundary between Caroline and Talbot counties.

Galloway Run rises in Precinct No. 3, Easton township, at altitude 50 feet above sea level and flows southeastward into Kings Creek (tributary to Choptank River, which flows into Chesapeake Bay) ; length, 1½ miles.

Geary Millpond; Queen Anne's and Caroline counties; receives the waters of Mason and Germantown branches of Tuckahoe Creek and of the Blackston Branch; discharges through Tuckahoe Creek to Choptank River (tributary to Chesapeake Bay). Denton sheet, U. S. Geol. Survey.

Glebe Creek rises in Easton township, at altitude 50 feet above sea level; flows westward into Miles River, one of the many tidal estuaries indenting the eastern shore of Chesapeake Bay in Talbot County; about 1½ miles of the head of this creek is creek-like in character.

Goldsboro Creek rises in Easton township; flows in general southwestward into Miles River, one of the tidal estuaries indenting the eastern shore of Chesapeake Bay in Talbot County.

Kings Creek, formed in Precinct No. 2, Chapel township, by the junction of Wootenaux Creek and Galloway Run; flows somewhat north of east 3 miles, then southeastward into Choptank River (tributary to Chesapeake Bay) near Kingston Landing.

Miles Creek rises in Trappe township, at altitude about 50 feet above sea level; flows southeastward 3 miles then northeastward 2 miles.into Choptank River, which discharges into Chesapeake Bay; lower course flows through a tidal marsh.

Mill Creek rises in the northern part of Chapel township, at altitude about 50 feet above sea level; flows southwestward 5½ miles into Skipton Creek, which discharges to Front Wye River, one of the tidal estuaries which cut the eastern shore of Chesapeake Bay in Talbot and Queen Anne's counties.

Norwich Creek, Queen Anne's and Talbot counties, rises near Starr in Queen Anne's County; flows southeastward into the northeastern part of Talbot County where it enters Tuckahoe Creek (tributary to Choptank River, which discharges to Chesapeake Bay).

Peachblossom Creek rises in Easton township, at altitude about 40 feet above sea level; flows westward into Tred Avon River, one of the many tidal estuaries in this part of Talbot County; only about 1½ miles of the head of the creek is creek-like in character.

Potts Mill Creek rises in Chapel township; northwest of Woodland, at altitude 50 feet above sea level; flows south of west into Miles River, one of the many tidal estuaries indenting the eastern shore of Chesapeake Bay in Talbot County; length of the creek to Miles River, about 4 miles.

Skipton Creek rises in Chapel township; flows in general northwestward into Front Wye River, one of the tidal estuaries which indent the eastern shore of Chesapeake Bay in Talbot and Queen Anne's counties.

Tuckahoe Creek, Queen Anne's and Talbot counties, formed in Queen Anne's County by Mason and German branches, which unite at altitude 20 feet above sea level, about 2 miles above Geary Millpond; flows in general southward into Choptank River, which discharges into Chesapeake Bay; marshy; forms boundary between

Talbot and Caroline counties. County maps, Queen Anne's and Talbot counties; Barclay and Denton sheets, U. S. Geol. Survey.

Williams Creek, a stream about 1½ miles long, rising in Easton township, Precinct No. 2, and flowing southeastward into Choptank River, which discharges into Chesapeake Bay.

Wootenaux Creek rises in Precinct No. 2, Chapel township, south of Woodland, at altitude about 55 feet above sea level; flows southerly into Kings Creek (tributary to Choptank River, which discharges to Chesapeake Bay) ; length, about 3½ miles.

THE MAGNETIC DECLINATION IN TALBOT COUNTY.

BY

LOUIS A. BAUER

INTRODUCTORY.

Values of the magnetic declination of the needle, or of the "variation of the compass," as observed by the Maryland Geological Survey and the United States Coast and Geodetic Survey at various points within the county are given in Table I.

For a general description of the methods and instruments used, references must be made to the "First Report upon Magnetic Work in Maryland" (Md. Geol. Survey, vol. I, pt. v, 1897). In the Second Report (Md. Geol. Survey, vol. V, pt. i, 1905) the various values collected were reduced to January 1, 1900. They are now also given for January 1, 1910 and 1925. The First Report gives likewise an historical account of the phenomena of the compass needle and discusses fully the difficulties encountered by the surveyor on account of the many fluctuations to which the compass needle is subject. To these reports the reader is referred for any additional details.

Table I. Magnetic Declinations in Talbot County.

TABLE I

Stations	Latitude N.	Longitude W. of Greenwich	Date when observed	Value observed	Magnetic Declination (west) Reduced to			Observer
					1900.0	1910.0	1925.0	
	° ′	° ′		° ′	° ′	° ′	° ′	
Oxford	38 41.4	76 10.5	1897.5	5 33.9	5 41	6 23	7 23	L. A. Bauer
Tilghman Island	38 42.9	76 20.0	1897.3	5 23.4	5 31	6 13	7 12	L. A. Bauer
Easton, Fair Grounds ..	38 46.0	76 04.4	1897.5	5 34.2	5 42	6 24	7 25	L. A. Bauer
Easton, west monument	38 46.5	76 05.0	1897.5	5 48.3*	5 56*	6 38*	7 39*	L. A. Bauer
Easton, west monument	38 46.5	76 05.0	1900.1	6 07*	6 07*	6 49*	7 50*	R. H. Blain, surveyor

Explanation: The date of observation is given in years and tenths of. January 1, 1900 would accordingly be expressed by 1900.0 and similarly for subsequent dates. See also Table II.

---

* Judging from Mr. Blain's letter of 1905, this site is no longer suitable, as confirmed also by information received by the United States Coast and Geodetic Survey in letter of March 14, 1925 from Messrs. Kastenhuber and Anderson, civil engineers and surveyors at Easton.

THE MERIDIAN LINE.

In compliance with the instructions from the County Commissioners July 15, 1897, L. A. Bauer, of the Maryland Geological
Survey, established a surveyor's true line on June 24, 1897, at the
Countyseat, Easton, on the Court House grounds. Approved astro-
nomical methods and instruments were used so that the line as
determined was correct within one minute.

Owing to the lay and character of the grounds around the Court
House it was found necessary to establish a true east and west line
instead of a true north and south line. By laying off 90 degrees the
surveyor can at any time obtain the true meridian. It was unfortu-
nate also that the instructions did not permit locating the line elsewhere because there was evidence of artificial local disturbances
near the Court House. In order to determine the approximate
amount of disturbance magnetic observations were also made on the
Fair Grounds (see description of stations).

The east and west line was marked by two granite posts. 7x7
inches square and 4½ feet long. They were imbedded in concrete
and firmly packed, each projecting about 5 inches above the ground.
A brass bolt one inch in diameter and 3 inches long, is leaded flush
in the center of each monument, the line passing through the centers
of the crosses cut on these bolts is the true east and west line. The
monuments are so placed that the letters N M on the east stone and
S M on the west stone would indicate approximately the direction
of the true meridian lines passing through the stones. The year
1897 occurs on each monument and the total length of the line is
201.4 feet.

In the official report supplied for the files of the Court House it
was recommended at the time (1897) that all compass observations
should be made over the west monument, the amount of disturbance
as compared with the observations in the Fair Grounds being 14
minutes to be subtracted (see Table I).

In accordance with information received from Mr. Robert H.
Blain, surveyor, dated Easton, May 1, 1905, it would appear that

owing to the erection of a large gas plant, containing considerable iron at about 200 feet southwest of the west monument, this site is no longer suitable for compass work. (See footnote Table I.)

DESCRIPTIONS OF STATIONS.

Oxford, 1897.—The station was on the beach in front of Sinclair's Hotel and is near the Coast and Geodetic Survey station of 1856.

Tilghman Island, 1897.—In Mr. B. B. Sinclair's field back of Mrs. Lee's hotel about 1 mile nearly north of the landing of the Baltimore, Chesapeake and Atlantic railway steamers; about 20 paces east of Mrs. Lee's back fence and the same distance south of the road on the north side of Mrs. Lee's grounds. The island is about one-half mile wide at this point and the station is just about midway. The steamboat landing is on the east side of the island.

Easton, 1897.—When establishing the meridian line in 1897, magnetic observations were made at various points. The station in the Fair Grounds outside the city is the one adopted in the Magnetic Survey. The station over the west monument in the Court House grounds is subject to artificial local disturbance. (See concluding paragraph under section Meridian Line.)

CHANGES IN MAGNETIC DECLINATION AT EASTON FROM 1700 TO 1925.

Table II has been reproduced from page 483 of the First Report without change except that it has been extended to 1925 with the aid of data supplied by the U. S. Coast and Geodetic Survey. It should be noted that the table refers to the Fair Grounds station.

TABLE II.

Year (Jan. 1)	Needle pointed	Year (Jan. 1)	Needle pointed	Year (Jan. 1)	Needle pointed	Year (Jan. 1)	Needle pointed	Year (Jan. 1)	Needle pointed
1700	5 49W	1750	2 58W	1800	1 02W	1850	2 38W	1900	42W
05	5 39W	55	2 40W	05	1 01W	55	2 56W	05	00W
10	5 27W	60	2 22W	10	1 04W	60	3 16W	10	24W
15	5 12W	65	2 06W	15	1 08W	65	3 36W	15	44W
20	4 56W	70	1 50W	20	1 15W	70	3 56W	20	04W
25	4 37W	75	1 36W	25	1 24W	75	4 15W	25	25W
30	4 17W	80	1 24W	30	1 36W	80	4 33W		
35	3 57W	85	1 15W	35	1 48W	85	4 52W		
40	3 37W	90	1 08W	40	2 04W	90	5 10W		
45	3 18W	95	1 05W	45	2 20W	95	5 26W		

The declination is west in the county and at present is increasing at the average annual rate of 4 minutes.

With the aid of the figures in Table II the surveyor can readily ascertain the amount of change of the needle between two dates. It will suffice for practical purposes to regard the amount of change thus derived as the same over the county. It should be emphasized, however, that when applying the quantities thus found in the re-running of old lines, the surveyor should not forget that the table cannot attempt to give the correction to be allowed on account of the error of the compass used in the original survey.

To reduce an observation of the magnetic declination to the mean value for the day of 24 hours, apply the quantities given in the table below with the sign as affixed:

Month	6A.M.	7	8	9	10	11	Noon	1	2	3	4	5	6P.M.
Jan.	—0.1	+0.2	+1.0	+2.1	+2.4	+1.2	—1.1	—2.5	—2.6	—2.1	—1.3	—0.2	+0.2
Feb.	+0.6	+0.7	+1.5	+1.9	+1.4	—0.1	—1.5	—2.1	—2.5	—2.0	—1.2	—0.8	—0.4
March	+1.2	+2.0	+3.0	+2.8	+1.6	—0.6	—2.5	—3.4	—3.7	—3.3	—2.3	—1.2	—0.5
April	+2.5	+3.1	+3.4	+2.6	+0.8	—2.1	—4.0	—4.1	—4.2	—3.6	—2.3	—1.2	—0.2
May	+3.0	+3.8	+3.9	+2.6	+0.1	—2.4	—4.0	—5.0	—4.5	—3.6	—2.3	—0.9	+0.1
June	+2.9	+4.4	+4.4	+3.3	+1.1	—2.0	—3.6	—4.5	—4.5	—3.8	—2.6	—1.2	—0.2
July	+3.1	+4.6	+4.9	+3.9	+1.8	—1.2	—3.4	—4.4	—4.7	—4.2	—2.8	—1.3	—0.3
Aug.	+2.9	+4.9	+5.4	+3.7	+0.4	—2.8	—4.7	—5.1	—4.9	—3.7	—1.9	—0.6	+0.3
Sept.	+1.8	+2.8	+3.4	+2.5	+0.3	—2.7	—4.4	—4.6	—4.2	—4.0	—1.4	—0.3	—0.1
Oct.	+0.5	+1.6	+3.1	+2.8	+1.4	—1.0	—2.7	—3.3	—3.4	—2.4	—1.3	—0.4	—0.4
Nov.	+0.5	+1.2	+1.7	+1.8	+1.1	—0.5	—2.0	—2.7	—2.6	—1.8	—1.0	—0.2	+0.2
Dec.	+0.2	+0.3	+0.8	+1.8	+1.8	0.0	—1.6	—2.4	—2.3	—1.8	—1.1	—0.3	+0.1

TRUE BEARINGS.

At West Monument:

True bearing of N.E. corner of Odd Fellows Hall, S65° 43'E.

True bearing of S.W. corner of Moreland Block, N36° 15'E.

The latitude of the Court House may be taken to be 38° 46'.5 and the longitude 76° 05'.0 W. of Greenwich. To obtain true local mean time, subtract from Eastern or Standard time, 4 minutes and 20 seconds.

THE FORESTS OF TALBOT COUNTY

BY

F. W. BESLEY

Introductory.

Talbot County lies midway between the typical hardwood forests of the northern Eastern Shore of Maryland and the prevailing pine types of the southern Eastern Shore section. In consequence there is a predominance of hardwoods in the northern part of the county, and a predominance of pine in the southern part, with all gradations of mixture of the two. Considered as a whole, 19 per cent of Talbot County is in pure hardwoods, 26 per cent in pine, and 55 per cent in mixed hardwood and pine. Some of this, particularly the pine, is in heavy stands, often running as high as 20,000 feet to the acre. Numerous waterways, two important railroad lines, and an extensive good road system, make all sections accessible and place all classes of forest products within easy reach of markets.

The area of the county is 171,520 acres, classified as follows:

Improved Farm Land	115,753 acres
Woodland	45,822 acres
Salt Marsh	3,392 acres
Waste Land	6,553 acres

The Distribution of Forest Land.

The forests are rather evenly distributed over the county, mostly in small holdings, although there are some large tracts in nearly every one of the election districts. The Easton District, No. 4, has the largest area of woodland and on the whole the woodlands are in larger bodies than in the other districts. 40 per cent is in woodland and the total stand of saw timber is considerably greater than that of any other district. It is also to be noted that there is about an equal amount of hardwood and of pine of saw timber size. St.

Michael's District, No. 2, has the smallest percentage of woodland of any district in the county, only 16 per cent, and the smallest wooded area, with the exception of District No. 5, Bay Hundred. There is a preponderance of pine over hardwood in the proportion of 9 to 1. There is a larger percentage of pure pine stands in this district than in any of the others. The forest areas are in smaller bodies than found in any of the other districts. In Trappe District, No. 3, 29 per cent of the land area is in forest and next to the Easton District it has the highest acreage of forest land. About two-thirds of the forest is pine and the remainder hardwood.

Chapel District, No. 4, while but 22 per cent wooded, ranks next to the Easton District in the amount of saw timber, of which about 40 per cent is hardwood. There are a few stands of pure pine in this district, but larger areas of purer hardwood.

Bay Hundred District, No. 5, consists of a long narrow peninsula extending southward from Claiborne, and is the smallest in area of any of the districts, and likewise has the smallest forested area, although the percentage of forest land, 19 per cent, is slightly greater than District No. 2 with 16 per cent. With the exception of some large bodies of mixed hardwood and pine in the northern part of this district, the stands are, practically, all pure pine. The proportion of pine saw timber to hardwood is about 10 to 1. The timber values, however, in this district are higher due to the heavy demands for building material and fire wood.

FOREST TYPES.

In mapping the forest land of the county in 1910, when a careful survey was made, the forests were divided into three main types,— hardwood, mixed hardwood and pine, and pure pine. The hardwood forests differ considerably in composition on different types of soil. On the low swampy ground of heavy soil, a preponderance of gum, maple, pin oak, and willow oak, while on the upland soils, white oak, black oak, Spanish oak, tulip poplar, and hickory are

FIG. 1.—LOBLOLLY PINE WIND BREAK NEAR WYE.

FIG. 2.—A GOOD STAND OF LOBLOLLY PINE NEAR TUNIS MILLS.

characteristic trees. On the higher, better drained soils, pure stands of pine occur. There are two species of pine, (*Pinus virginiana*) spruce pine, which occurs sparingly in the eastern and southeastern sections of the county, and the loblolly pine, (*Pinus taeda*) which is the predominant tree throughout the county. The spruce pine occupies the higher, dryer, and poorer type soils, while the loblolly pine is found in the moister but better types of the lighter soils.

This species is by far the most important timber tree in the county, and is cut more extensively than any other species. Due to the close cutting and competition with the vigorous hardwood growth, it is less abundant in the second growth forests than was the case 50 years ago. The mixed hardwood and pine type occupies a much larger acreage than any other, due largely to the excessive cutting of the pine, thus opening the stands to the introduction of the more persistent hardwoods. The mixed hardwood and pine type is therefore increasing while the pure pine type is decreasing. Fires have also helped to bring about this change. A fire in the woods will completely kill the small pine trees, while the hardwoods which may be burned to the ground will sprout up from the roots, giving them an advantage over the pine. In this way a natural pine forest may be converted to a hardwood forest, or at least to a mixed hardwood pine forest.

WOODED AREA, STAND AND VALUE OF SAW TIMBER BY ELECTION DISTRICTS

Dist. No.	Total Land Area	Wooded Area Acres	Per cent Wooded	Stand of Saw Timber in Board Feet			Stumpage Value		Total
				Hardwood M Bd. Ft.	Pine M Bd. Ft.	Total M Bd. Ft.	Hardwood $4.00 Per M	Pine $10.00 Per M	
1	35,680	14,369	40	52,795	45,247	98,042	$422,360	$452,470	$874,830
2	22,920	3,699	16	1,734	18,620	20,354	13,872	186,200	200,072
3	46,720	13,613	29	10,598	23,382	34,980	84,784	233,820	318,604
4	54,735	11,955	22	20,186	32,829	53,015	161,488	328,290	489,778
5	11,465	2,186	19	557	7,292	7,849	4,456	72,920	77,376
The County...	171,520	45,822	27	85,870	127,370	214,240	$686,960	$1,273,700	$1,960,660

NATIVE AND INTRODUCED TREES.

CONIFERS.

Scientific Name	Common Name
Chamaecyparis thyoides L	Southern White Cedai
Juniperus virginiana L	Red Cedar
Pinus echinata Mill	Short Leaf Pine
Pinus rigida Mill	Black Pine
Pinus taeda L	Loblolly Pine
Pinus virginiana Mill	Spruce Pine

BROAD LEAVES.

Acer negundo L	Ash-Leaved Maple
Acer rubrum L	Red Maple
Acer saccharinum L	Silver Maple
Ailanthus altissima Swins	Ailanthus
Alnus maritima Nutt	Swamp Alder
Amelanchier canadensis Med	Service Berry
Aralia spinosa L	Hercules Club
Asimina triloba Dunal	Paw Paw
Betula nigra L	River Birch
Carpinus caroliniana Walt	Blue Beech
Castanea dentata Borkh	Chestnut
Celtis occidentalis L	Hackberry
Cercis canadensis L	Red Bud
Cornus alternifolia L	Dogwood
Cornus florida L	Flowering Dogwood
Diospyros virginiana L	Persimmon
Fagus gradifolia Ehrh	Beech
Fraxinus americana L	White Ash
Fraxinus pennsylvanica Marsh	Red Ash
Gleditsia triacanthos L	Honey Locust
Hamamelis virginiana L	Witch Hazel
Hicoria alba L	White Hickory
Hicoria glabra Mill	Pignut Hickory
Hicoria minima Britt	Butternut Hickory
Ilex opaca Ait	Holly
Juglans cinerea L	White Walnut
Juglans nigra L	Black Walnut
Liquidambar styraciflua L	Red Gum
Liriodendron tulipifera L	Tulip Poplar
Magnolia virginiana L	Sweet Bay
Morus rubra L	Red Mulberry
Myrica cerifera L	Wax Myrtle
Nyssa sylvatica Marsh	Sour Gum

Scientific Name	Common Name
Paulownia tomentosa Stend............Empress Tree
Platanus occidentalis L................Sycamore
Populus alba L.......................Silver Poplar
Populus nigra italica ....Lombardy Poplar
Prunus americana Marsh...............Wild Plum
Prunus pennsylvanica L...............Fire Cherry
Prunus serotina Ehrh.................Wild Black Cherry
Prunus virginiana L..................Choke Cherry
Quercus alba L.......................White Oak
Quercus coccinea Muench..............Scarlet Oak
Quercus imbricaria Michx.............Shingle Oak
Quercus lyrata Walt..................Overcup Oak
Quercus marilandica Muench...........Black Jack Oak
Quercus michauxi L...................Basket Oak
Quercus stellata Wang................Post Oak
Quercus nigra L......................Water Oak
Quercus palustris Muench.............Pin Oak
Quercus phellos L....................Willow Oak
Quercus bicolor Willd................Swamp White Oak
Quercus montana L....................Chestnut Oak
Quercus falcata Michx................Swamp Red Oak
Quercus velutina Lam.................Black Oak
Robinia pseudocacia L................Black Locust
Salix alba var. *vitelina*............White Willow
Salix nigra Marsh....................Black Willow
Sambucus canadensis L................Elder
Sassafras sassafras Karst............Sassafras
Toxylon pomiferum Rafn...............Osage Orange
Ulmus americana L....................White Elm
Ulmus fulva Michx....................Slippery Elm

Talbot is a county noted for big trees. The favorable conditions of soil and climate produce large individual trees, as well as good forests. In the list of noted trees, the Wye Oak, a giant white oak overhanging the state highway at Wye Mills, stands supreme. Dr. Sargent, the greatest authority on trees in this or any other county said when he saw the Wye Oak that it had the largest base measurement and the greatest spread of any oak that he had seen in America.

Other trees of particular note, because they are the largest of their kind of the 450 entries submitted to the Maryland Forestry

Association in a state-wide Big Tree Contest conducted in 1925-1926 are a beech, a red cedar, and English yew, and a weeping willow at Hope House, the residence of Mrs. Wm. J. Starr; a willow oak and a chestnut oak at Myrtle Grove; and a mimosa tree at G. M. Tull's place in Oxford. There are, probably, other "biggest" trees in Talbot County that have not yet been discovered.

IMPORTANT TIMBER TREES.

Loblolly pine stands pre-eminent as the most important timber tree of the county, and the best stands of timber are those of pure loblolly pine. Other species, such as white, black, and Spanish oak have a higher value per thousand feet. Their much slower growth and the large amount of waste in cutting make these species, however, much less valuable for timber growing. Loblolly pine will grow rapidly on the poorer soils and produce a very much larger volume of merchantable timber per acre than any other species. Furthermore, this particular tree has such a variety of uses as to insure a certain and steady demand.

Its principal uses are lumber in general construction; the making of boxes and crates; staves and headings, which is an important local use; the making of barrels for potatoes and other farm products; mine props which take trees with a middle diameter of from 7 to 16 inches; piles, which require straight stems from 12 to 14 inches in diameter at the butt, and lengths up to 80 feet; pulpwood, sold by the cord using small size trees down to as low as 3 inches in diameter inside the bark at the small end and cut in 5 foot lengths; down to the very large use for fuelwood, especially in the sections removed from towns and particularly where hardwood is not easily obtained.

Of the oaks found in the county, white oak is the most valuable, and is much in demand for bridge plank, general construction, and railroad ties. The other oaks, including black, spanish, pin, and willow oak are cut into framing, dimension stock, and railroad ties,

for which the demand far exceeds the supply. Red gum is dis-
tinctively a swamp species used principally in the smaller sizes and
for veneer logs in diameters of 18 inches and up. Gum veneer is
used principally in the making of baskets.

LUMBER AND TIMBER CUT.

There are 21 saw mills operating in the county with a total
lumber cut for the year 1925 of about 2,000,000 feet, board measure.
Nearly all of the mills are of the portable type, moving from place
to place wherever stumpage can be obtained. A few of them are
stationary, doing custom sawing, but their output is small as compared with the others. About 70 per cent of the lumber cut is pine,
the remainder hardwoods, principally oak, poplar, and gum. Nearly
all of the lumber cut is used locally for farm buildings, bridges, and
miscellaneous construction,—only a small quantity is shipped out
in the form of sawed ties and car stock. More lumber is imported
than is produced locally, bringing the annual consumption to more
than 5,000,000 feet.

Other forest products are veneer logs, of which 150,000 feet was
shipped in 1925, and a small quantity of piling, mine props, and
pulpwood. Home needs for fish poles, fence posts, and fuel wood
amount in the aggregate to a large quantity which is difficult to
estimate. In cubic foot content, it far exceeds all other uses combined. A conservative estimate is 150,000 cubic feet of wood products taken from the forests annually, which is considerably in excess
of the annual growth. This means that the forest capital is being
depleted, and unless growth can be increased the annual cut will
fall off rapidly. In fact, it has fallen off greatly in the past 20 years,
and few stands of large sized timber are left. It is confidently believed, however, that stopping forest fires, all of which are pre-
ventable, and the handling of the forests in accord with correct
forestry principles would in ten years, on the present forest area,
increase timber growth 25 per cent and in 20 years 50 per cent over

present yields, going a long way toward making the county self supporting so far as wood products are concerned.

Destructive Agencies.

Constant changes are being brought about in the forests, due to a combination of agencies. Excessive cutting, particularly the pine and the more valuable species of hardwoods, leaving uncut the less valuable hardwood, has greatly disturbed the 'balance" and brought about a steady deterioration, not only in the composition of the forests, but also in their productivity. The "balance" has been disturbed in that the better species have been repeatedly cut out, while the poorer trees have been undisturbed, and thereby had a monopoly of the light and growing space. This was not by design, but due to economic causes, chief of which was that it did not pay years ago to take out anything but the best, consequently in the mixed type of forest, consisting of pine and hardwoods, the pine and the best of the hardwoods have suffered by close cutting, while the less valuable maple, elm, and other low grade species have taken advantage of the openings caused by the cutting of other trees, and each successive cutting has led to further deterioration.

Forest fires have been less destructive in Talbot County than in most counties of the state, but there have been extremely dry seasons when a great deal of damage has been done. Forest fires do more damage to pine forests than to hardwood, especially during their early life. When a young pine forest burns, there is not only complete destruction of the trees, but it often takes many years to reseed and restock the stand, while in the case of young hardwoods, although the trees themselves may be burned, they almost invariably send up shoots from the roots and the next season after the fire a new growth starts. The sprouting capacity of the hardwoods, with the inability of pines to grow again, gives the hardwoods the advantage in a mixed forest, and accounts in a considerable measure for the disappearance of pine and increase in the hardwoods. Not only do forest fires destroy the young forests, but they seriously damage the larger trees, producing scars, rotted butts, and general deterioration in the quality of timber, as well as impoverishment of the soil and the retarding of the growth.

The State Department of Forestry has a forest protection organization, embracing all the counties of the state, represented by the Forest Wardens, of which there are several located in Talbot County. These men have full authority to employ or summon help to suppress fires. Their names and telephone numbers will be found in the back of the local telephone directory. The nearest warden should be promptly notified in case of a forest fire.

INSECT DAMAGE.

There are a number of insects that do some damage to forest trees, but the damage in most cases is very slight as natural enemies keep the insects well under control. The most serious damage to the loblolly pine that has developed in the last few years is from the bud moth (*Pinipestis amatella*). This is a small insect, the larvae of which eats out the center of the bud after it is formed in late summer. Growth of the succeeding seasons must then start from a bud developed lower down on the shoot, not only retarding the development of the trees, but causing a crook in the stem. The tree is usually attacked when from 2 to 4 feet in height. The vigor of growth of this species usually enables it to overcome these successive attacks, and after it reaches the height of about 5 feet, it appears to be immune to further damage.

The pine bark bettle (*Dendroctonus frontalis*), another insect attacking loblolly pine, has been noted in the county doing some damage in isolated areas. This beetle brings about the death of the trees by tunnelling and feeding on the inner bark of the tree in such vast numbers that the tree is actually girdled and the food supply cut off. The infested tree should be cut and promptly utilized, or the bark removed and burned, especially along the lower part of the trunk. It is easy to pick out the infested tree by the fact that the needles turn sickly yellow, then brown. The insects are apt to con- regate in a single tree, and by cutting and burning such trees, most of the insects are destroyed.

THE FUTURE OF THE FORESTS.

The rapid increase in timber values and the presence of large areas of forest land in the county emphasize the need of the appli- cation of the principles of forestry in the growing of timber. The use of this land for the growing of timber crops is just as important as the growing of agricultural crops on other land. There is also as great an opportunity for the improvement of condition for the grow- ing of timber as there is in the improvement of field crops. In order to arrive at a correct understanding of the problem, certain basic facts and principles should be recognized: First, the timber sup- plies of the country are being rapidly depleted and since timber is one of the indispensable products and one that is bulky to ship, local needs must be provided for. Second, there is probably no section of the country more favorable for timber growing, due to species of the most rapid growth and highest value, favorable soil and cli- mate, excellent transportation system by water, rail, and good roads, nearness to the consuming centers, and general freedom from destructive insect and tree diseases.

In the first place, the forest must be recognized as a growing crop. As such, the most favorable conditions for the highest yield of the highest value must be brought about. Since loblolly pine is the most profitable tree to grow on soils for which it is adapted, this species should be encouraged above all others, except in the swamp lands of heavy clay soil. The keen competition of hardwoods, especially in the young stands, makes it often necessary to clean out the competing hardwoods to give the young pines a chance. Once released, their rapidity of growth will usually insure ultimate su- premacy. In the case of a natural stand of pure pine, a new pine

FIG. 1.—VIEW IN 13-YEAR-OLD CATALPA PLANTATION AT HOPE HOUSE.

fórest may be re-established after cutting by leaving 3 or 4 seed trees to the acre, unless there is a heavy undergrowth of hardwoods. In the latter case, cutting the hardwoods closely and burning off the land before the mature trees are cut, especially just prior to a good seed year, will often insure a reseeding of the pine.

On swampy land, the red gum, often called sweet gum, or white gum, together with the pin oak and willow oak, are trees to be favored, and in making cuttings for pulpwood or other purposes, the less valuable species should be cut and utilized in order to give these more valuable species a better chance to develop.

FOREST PLANTING.

There is waste land on nearly every farm that would bring better returns planted in forest than in any other crop. Where lands are badly gullied, the planting of locust is advised, as this will not only hold the soil, but produce a valuable crop of durable fence posts in less time than any other species. There are other areas of poor land, not bringing fair returns in field crops that could be better utilized in the growing of loblolly pine. Many demonstration plantings have been made in the county in the past few years, most of them succeeding to a remarkable degree. There is need of planting pines for windbreaks to protect houses and farm buildings from the sweep of the northwest winds. For this purpose loblolly pine, Scotch pine, or Norway spruce are particularly well adapted.[1]

THE STARR ARBORETUM.

Mrs. Wm. J. Starr, the owner of one of the oldest and most beautiful estates, known as Hope, has brought together a large number of trees, both native and foreign, and purposes establishing an arboretum, where all the important native and foreign species, adapted for planting on the Eastern Shore may be seen in their natural form.

---

[1] The State Department of Forestry maintains a forest nursery from which trees are distributed to land owners in the state at cost, upon application.

The collection is being constantly augmented by new species, each suitably labeled. The State Department of Forestry is co-operating with Mrs. Starr in furnishing specimens of all tree species native to the state. In a few years, when the arboretum specimens have had time to develop, this place should be a real mecca for those who love trees and want to become better acquainted with them.

INDEX